APPLICATION OF WALLS TO LANDSLIDE CONTROL PROBLEMS

Proceedings of two sessions
sponsored by the
Committee on Earth Retaining Structures
of the
Geotechnical Engineering Division
of the
American Society of Civil Engineers

at the
ASCE National Convention
Las Vegas, Nevada
April 29, 1982

R. B. Reeves, Editor

Published by the
American Society of Civil Engineers
345 East 47th Street
New York, New York 10017

Library of Congress Catalog Card No. 82-70766
ISBN 0-87262-302-5
Manufactured in the United States of America.

PREFACE

Once we were able to expand our urban frontier at will; abundant land allowed us to leave conjested areas, avoid undesireable land and alter the landscape without thought to negative consequences. Now commerce and industry must develop by a more frugal approach that is based on dense population and limited lands. Engineers must satify these needs with inovative solutions that utilize new technology. These papers are presented with this in mind.

Landslides cause loss of life and property when they occur in densly populated or intensly developed regions. For these reasons landslides have become more severe in North America with the passage of time. As a corollary, the established slide control methods of avoidance, drainage and regrading are not always practical. Walls have been built to stabilize slopes, and often the results have been far less than expected. Tiebacks were introduced to North American practice from Europe in the late 1960's, and soon engineers found them ideally suited to stabilizing certain landslides.

Regrading, drainage, buttresses, cribs and earth reinforcement are slide control measures which have been thoroughly discussed in the literature. The literature does not similarily describe tieback practice; hence, we have prepared these papers to fill this void. Each author has experience with his subject, and all authors combine to give a broad point of view. This is the view of current practice (1982) by engineers who have met successfully the requirements for stabilizing specific slides.

The four case histories together present the evolution of tieback restraint systems and separately describe how the authors solved a specific problem. Engineering geology, stability analysis and shear strength, the triad upon which each case history is based, are discussed on a conceptual basis because there is no single theory for an applied art that must be learned by field experience. Current accepted practice for tieback design and construction has been prepared by authors who are personally familiar with the practice that they advocate.

Our knowledge of the use of walls to control slides is far from complete. Each new application carries uncertainity and concern; however, each new application is a problem that some engineer has to solve. To us who have to meet this challange we direct this special publication. Our objectives are to increase your confidence in using tieback walls for certain slide control applications and to expand knowledge of this subject through your critical review of these papers.

CONTENTS

GEOLOGIC ASPECTS OF LANDSLIDE CONTROL USING WALLS

by Robert L. Schuster,[1] F. ASCE, and Robert W. Fleming[2]

INTRODUCTION

The term "landslide" traditionally has been defined as the downward and outward movement of slope-forming materials--natural rock, soils, artificial fills, or combinations of these materials (37). In theory, landslides constitute the group of slope movements wherein a shear failure occurs along a specific surface or combination of surfaces; thus, strictly speaking, it does not apply to all types of slope movements (38). However, because of its common usage, particularly in the civil engineering literature, the authors have decided to follow common practice and use landslides as the collector term for the various types of slope failures.

It is estimated that costs of damages from landslides in the United States exceed $1 billion annually (27). This total is supported by more detailed studies of smaller geographic areas. For example, a recent survey of costs of landslide damage in San Diego County during the rainy seasons of 1978-79 and 1979-80 documented 120 landslides that caused damages of about $19 million (29). Eighty-three of these landslides caused damage to private property at an average cost per residence of $85,000, a value that equaled or exceeded the average property value. In light of this huge cost of landslides, the utilization of successful and economical remedial measures for prevention and control of landslide problems is important.

Terzaghi (35) has summarized the processes leading to landslides in his classic paper on the mechanisms of landsliding. In general, he divided causative factors into those conditions that exist in a slope, such as topography, lithology, and structural features, and those conditions that may produce a change, such as excavation, seismic events, and increase or removal of water. Selection of prevention or control measures for a potential or existing landslide requires an understanding of the landslide process, which, in turn, demands a knowledge of the local geology and the properties of the earth materials involved.

Landslide remedial measures fall into four general categories: relocation (i.e., avoidance), drainage, removal (modification of slope), and restraint (26). If relocation to avoid a landslide is possible, it is probably the surest way to prevent damage and should never be overlooked. It is applicable to all types of landslides, and in many cases proves to be less expensive than construction or excavation. However, it often is not cost-effective to relocate a major construction project. In such cases, stabilization methods must be used, and drainage and modification-of-slope measures are probably the most common.

---

[1]Civil Engineer/Geologist, [2]Geologist, Engineering Geology Branch, U.S. Geological Survey, Box 25046, MS 903, Denver, Colorado 80225

Where these methods do not suffice, however, artificial restraining barriers, such as buttresses, retaining walls, or rows of piles across the path of the landslide, commonly are used. These restraining structures in most cases are applied in conjunction with drainage and/or modification of slope. Properly engineered restraining structures have a useful role, particularly where lack of space restricts the effective use of slope modification to develop the slope to full design length. However, the use of restraining structures should be limited to prevention or control of small-scale landslides because they seldom are completely effective on larger ones (2).

The most commonly used restraining structures for slope stabilization have been gravity or cantilever retaining walls placed with their bases somewhat below the potential failure surface. In recent years, anchored walls or wall systems consisting of rows of large diameter reinforced concrete cylinders have been used to carry forces generated by slope movement to a line of reaction below the failure surface (9). All walls require adequate drainage to prevent buildup of water; free-draining landslide materials have been effectively controlled with permeable walls.

This paper will deal with geologic aspects of the use of walls as restraining structures, noting the types of landslides to which they are most applicable as well as their limitations as remedial measures. Veder (40) presented case examples of different types of walls in different physical conditions. Schweizer and Wright (28) noted many case histories of the use of walls to restrain landslides; some walls were successful, some were not. Instances of failure of retaining walls are common in civil engineering literature. As an early example, Ladd (18) noted the failure of eight retaining walls of various types due to small landslides in embankments and sedimentary rocks in the Appalachian Mountains. Skempton (30) discussed the causes of failure of five walls supporting slopes cut into stiff, fissured London clay. Gould (9, p. 261) commented on some commonly encountered conditions leading to possible failure as follows:

> "There are several potential difficulties with the conventional retaining structure. For one thing, excavation in front of a wall can reduce overall slope stability more than it is increased by the presence of the wall. If this is not ultimately the case, the necessary construction excavation can cause a temporary decrease of slope stability. Even though piles are used, a failure surface could develop at some depth below the base of [the] wall applying displacing rather than resisting soil forces on the upper portion of the piles and putting both vertical and batter piles in bending. Thus, if a conventional retaining structure is planned, consideration should be given to the change in overall slope stability occasioned by excavation in front of the wall. Among the most treacherous materials are heavily overconsolidated clays or clay-shales that are characterized by strain softening and possibly by high horizontal stresses inherited from their preloading. The worst possible conditions might be expected in an old slide area where movement has decreased strength towards residual values. In such

> materials the effect on the balance of forces of excavation for a conventional wall must be viewed most conservatively."

In addition to creating stress concentrations and changing the balance of forces, excavations create changes in subsurface hydrology that can adversely affect stability. Relaxation of a slope following excavation causes an increase in hydraulic conductivity that can lead to saturation and failure.

TYPES AND CHARACTERISTICS OF SLOPE MOVEMENTS

Classification

In the last 25 years, more than two dozen partial or complete classifications of landslide types have been published in various languages. The one most commonly used in North America is that of Varnes (38), which is based primarily on type of movement and secondarily on type of material involved. Types of movement have been divided by Varnes into five main groups: *falls*, *topples*, *slides*, *spreads*, and *flows* (Figure 1). A sixth, and very common, group can be called *complex* slope movements; this group includes combinations of two or more of the five primary groups. Typical diagrammatic examples of these groups are shown in Figure 2. Varnes has divided the materials in which landslides occur into two classes: rock and engineering soil; engineering soil is subdivided into debris and earth, coarse- and fine-grained categories, respectively.

TYPE OF MOVEMENT			TYPE OF MATERIAL		
			BEDROCK	ENGINEERING SOILS	
				Predominantly coarse grained	Predominantly fine grained
FALLS			Rock fall (Rock fall avalanche)	Debris fall	Earth fall
TOPPLES			Rock topple	Debris topple	Earth topple
SLIDES	ROTATIONAL	FEW UNITS	Rock slump	Debris slump	Earth slump
	TRANSLATIONAL		Rock block slide	Debris block slide	Earth block slide
		MANY UNITS	Rock slide	Debris slide	Earth slide
LATERAL SPREADS			Rock spread	Debris spread	Earth spread
FLOWS			Rock flow	Debris flow (Debris avalanche)	Earth flow (Mud flow)
COMPLEX			Combination of two or more types of movement		

Figure 1.--Classification of slope movements (after reference 38).

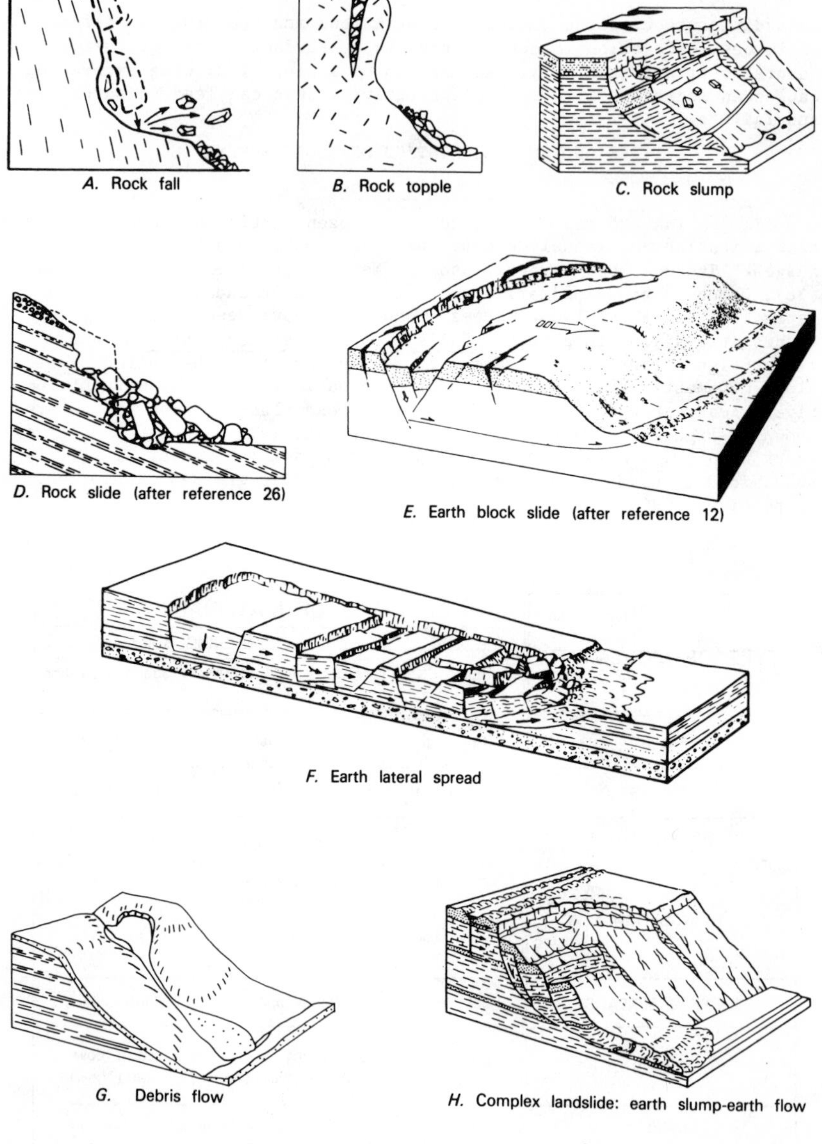

Figure 2.--Examples of common slope movement types (A, C, F, and H after reference 38).

General Characteristics

Landslides vary remarkably in occurrence and characteristics, ranging from small slumps to enormous masses involving cubic kilometers of earth material. They range in sense of displacement from vertical to nearly horizontal, in shape from sheets and slabs to blocks, wedges, and tongues, and in duration of activity from seconds to years (16).

Of considerable interest in planning and design of retaining walls is the rate of movement of a landslide. Landslide velocity can range from a few millimeters per year for creep to meters per second for falls and avalanches; Figure 3 presents ranges of general rates of movement for examples of the common types of landslides. For landslides in both soil and rock the maximum velocity is dependent on (a) steepness of slope, (b) shape of the failure surface, and (c) the physical properties of the earth materials. The greatest velocities occur in falls and avalanches on very steep slopes. Translational slides generally have greater velocities than slumps.

The highest velocities occur in rocks and soils that have low residual shear strength in comparison with peak strength (3). In addition, landslides occurring in soils containing an abundance of water commonly have greater velocities than do those in unsaturated soils; especially bad are quick clays and loose saturated sands and silts that are subject to liquefaction. However, dry rock fall avalanches and flows of dry sand or silt on steep slopes can also attain high velocities.

Where the cause of potential instability is erosion, particularly

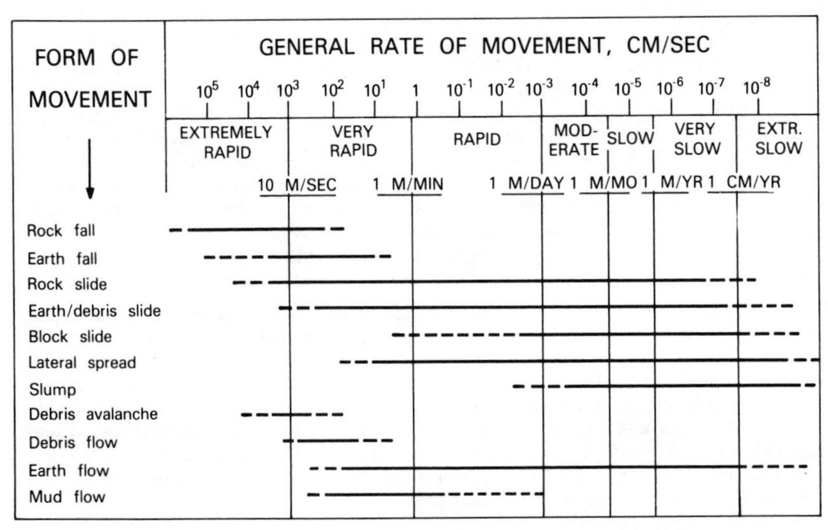

Figure 3.--Ranges of rates of movement for landslides. Dashed portions of horizontal bars represent relatively uncommon or poorly documented occurrences. (Modified from reference 16.)

erosion at the toes of slopes due to river or wave action, toe walls have been successful in maintaining stability. Ward (42) noted an example where prevention of wave erosion of the toe of sea cliffs at Folkestone Warren, England, enabled stabilization of the slope against large-scale slipping and rock fall. Similarly, Natarajan et al. (21) demonstrated the effectiveness of a concrete toe wall in protecting a riverbank against erosion; this action, in conjunction with the use of other remedial measures, increased the stability of the area upslope.

GEOLOGICAL CONTRIBUTIONS TO LANDSLIDE CONTROL

In general, the contributions of geologic studies to landslides fall into three separate but related areas:

(a) The first area of contribution is based on the widely accepted principle of geology called the "Principle of Uniformitarianism." This principle holds that the present is the key to the past and that study of present earth materials and processes can lead to an understanding of processes that have operated in the past to create materials and landscapes. For landslide studies, the principle is reversed so that the past is the key to the present and future. Detailed studies of ancient and modern landslides can lead to predictions about future landslide movements. For the most part, the predictions are conveyed in the form of maps. These maps commonly show the locations of previously failed slopes and perhaps the materials involved in the failures. Inferences can be made about the type of movement and the size, rate, and age of the failed slopes. The inherent causes of the failures can be evaluated by combining basic geologic information such as lithology, structure, surficial deposits, and ground-water conditions with the information on past landsliding.

Such geologic studies can result in a "zonation" of an area in terms of potential or actual hazard from landsliding. Varnes et al. (39) have prepared a comprehensive review of the principles and practice of zonation. By studying a well-made zonation map, a reader can ascertain not only the relative or quantitative hazard from potential landslide movements, but also the physical attributes that have been combined to produce the zonation ranking, such as dip of bedding, geologic materials, failure history of a slope, etc. Studies at locations where geological situations are similar to those existing at a proposed construction site may reveal the most likely type of failure and its characteristics, and the potentially most successful remedial measures.

(b) Geologists also provide a broad areal perspective of conditions and processes that is invaluable to landslide studies. Areas of previously failed slopes are commonly much larger than an engineering site such as a residential lot or subdivision. Subsurface hydrologic conditions may depend on sources of infiltration and discharge that are miles removed from an engineering site. A framework for the three-dimensional distribution of materials at a site can be inferred from a knowledge of the geologic history and, thus, can lead to an intelligent program of site exploration and characterization.

(c) And finally, with respect to conditions at a site, geological studies should provide physical and historical descriptions of the materials that are useful in the assessments of local stability. Physical descriptions of subsurface hydrology, locations and attitudes of weak materials or zones, and inherent variability of the various materials are important in assessing potential for different types of failure. Historical descriptions reveal the genesis of the deposits and history following deposition or formation of the materials. Such information provides a basis to anticipate, for example, the consolidation characteristics of clays and explanations for the three-dimensional configurations of the materials.

RELATIONSHIP OF CHARACTERISTICS OF TYPES OF LANDSLIDES TO SUCCESS OF RETAINING WALLS

Falls

Falls are most common in jointed rock on very steep to vertical or overhanging slopes. They are caused by erosion, weathering, frost action, temperature differentials, or water pressures in joints or other fractures. Usually the volumes of individual rock falls dealt with in engineering projects are small. The commonest case in the United States involves small- to medium-size rocks falling or bounding down cut slopes for highways or railways. However, exceptional rock fall avalanches with volumes in the range of millions of cubic yards, or larger, have occurred. Some of these, such as the 1881 Elm, Switzerland, and the 1970 Huascaran, Peru, avalanches are among the largest and most destructive landslides recorded (41). Rock fall avalanches attain high velocities, and, even though they begin on steep slopes, they commonly travel great distances in their runout onto gentle slopes.

Earth falls often occur where an erosion-resistant soil, such as an overconsolidated clay, is underlain by an erodible soil, such as a layer of silt or sand (32). The mechanism of earth falls in such layered soil systems has been described by Henkel (13). In heavily overconsolidated clays, falls often occur where rainwater or meltwater fills tension cracks at the crest of a steep slope (3).

Retaining walls are seldom used to prevent rock or earth falls. More commonly, controlled scaling or trimming of loose or unstable rock or soil from steep slopes or bolting loose blocks of rock to the stable mass is used. However, walls quite often are used to control rock that has already fallen by serving to prevent rocks from moving far enough to cause damage. Catch walls placed at the foot of a steep rock or rock-and-soil slope have been used to prevent rocks up to a size of about 6 ft (2 m) from rolling or bouncing onto a highway or railway. In recent years steel mesh nets and wire fences have replaced the walls for this purpose in most instances.

Before designing a catch wall, it is important to have some conception of the character of the expected rock fall. Rocks can reach the structure by falling free, bouncing, rolling, or sliding down the slope (23). The manner of their arrival at the wall depends on their size and shape, the character of the surface on which the rocks are moving, and particularly on the angle of slope (Figure 4). Other factors of importance are surface character of slope, height of source cliffs above the slope, broken slope features, kind of rock, gravity, and time (25).

Sliding or rolling rocks are easiest to intercept because their movement is in contact with the slope. For free-falling rocks, means of protection include preventing the fall or avoiding the area. For

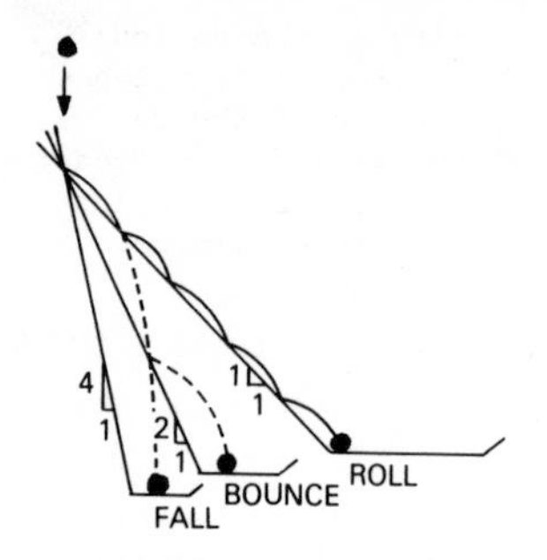

Figure 4.--Predicted path of falling rock for various slope angles (after references 24, 25).

transportation routes, the roadway can be covered with a rock shed or tunnel or the rocks kept from bouncing or rolling by means of a catchment structure. Design of catch walls for bouncing rocks is complicated by the difficulty of anticipating their trajectories. Piteau has devised a computer model for predicting the statistical probability of rocks arriving at various distances from the toes of unstable rock slopes (24). We hope that further work will allow prediction of heights of bounce for use in design.

The wall must be a stable structure, anchored to rock if necessary. Where the rolling or bouncing rock is expected to be large and angular, and to have high kinetic energy, even massive concrete walls can sustain severe damage (23). The French Laborataire Central des Ponts et Chaussees has noted several examples of successful wall designs for this purpose (10); included are walls of concrete blocks precast in sections and linked or keyed to prevent overturning, various types of crib walls, and gabion walls.

Topples

Topples have been recognized only recently as a distinct type of slope failure. As shown in Figure 2B, toppling consists of failure by forward rotation of an earth block or mass under the action of gravity and forces exerted by adjacent blocks or masses or by fluids or ice in cracks (38). Topples are most commonly found in jointed rock or rock with steeply inclined bedding comprising steep slopes or near-vertical cliffs. Examples of rock topples in Great Britain described by de Freitas and Watters (6) range in volume from 130 yd^3 (100 m^3) to 1.8X10^9 yd^3 (1.4 Gm^3). The toppling process develops slowly through an initial rotation of a few degrees, but when it actually reaches the failure state, the toppling mass moves very rapidly down the slope in much the same manner as a rock fall or rock slide. In fact, toppling failures commonly lead to falling or sliding, the exact mode of culmination depending on size, on shape of the toppled blocks, and on the character of slope below the topple.

Successful prevention of toppling can best be accomplished by rock bolting or the use of anchored cables or cable-net restraining devices. However, if a toppling failure occurs, retaining walls may have a function in protection of downslope structures, facilities, or people from the falling or sliding earth mass. In planning walls for such control, the same concepts should be used as for falls or slides.

Slides

Slides can occur in either rock or soil; displacement occurs along one or more closely spaced failure surfaces. Failure may be progressive: instead of failure occurring simultaneously, an overstressed zone can occur at a local point in the slope and progress to other points, thus forming a failure surface or surfaces.

As shown in Figure 1, slides can be divided into rotational and translational movements of rock and/or soil. The characteristics of these two types of slides and their relationships to retaining walls will be discussed separately. A large amount of detail on this subject is available in the literature; we will be able to treat it only cursorily here.

Rotational Slides

Although they occasionally are found in hard rock, rotational slides characteristically occur in fairly homogeneous clays or shales as slumps with failure surfaces that are concave upward; movement is generally rotational about an axis parallel to the slope. Slumps in these materials are relatively deep seated, with depth-to-length ratios ranging from 0.15 to 0.33 (31); there is a tendency for the higher values of this ratio to be associated with steeper slopes.

Classic purely rotational slope failures (slumps) most commonly occur in relatively homogeneous materials such as those found in constructed fills and embankments; this is one of the reasons for the interest of engineers in this mode of failure (38). Geologic materials are seldom uniform, however, and the shapes of natural slides are controlled by joints, faults, stratigraphy, and other discontinuities. The effects of stratification on the shape of the failure surface are illustrated in Figure 5. A noncircular failure surface results in distortion of the failed mass and the development of additional shear failures within the mass.

As shown in Figure 3, the velocity of slumps is relatively low. Thus velocity is not generally a problem in using walls to retain slumps. However, where slumps degenerate into flows, conventional retaining walls generally will not be successful.

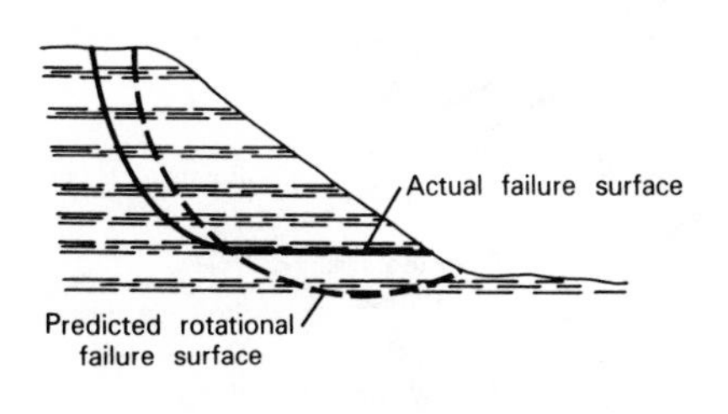

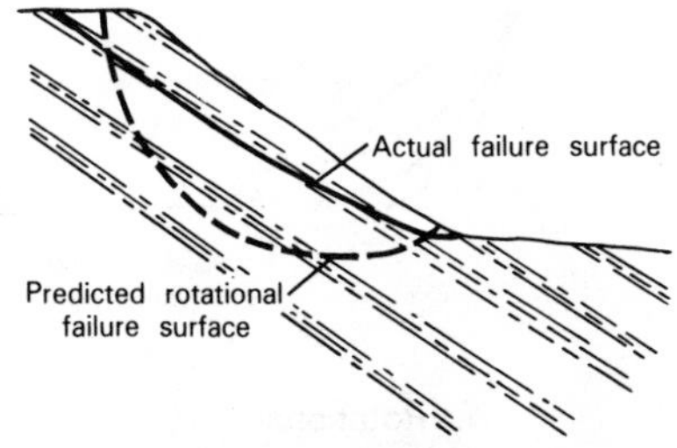

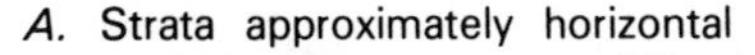

A. Strata approximately horizontal   *B.* Strata dipping downslope

Figure 5.--Effect of stratification on shape of failure surface.

If a rotational failure occurs along a surface of sliding that intersects the slope at or above its toe, the slide is known as a slope failure (36, p. 233). If, however, the earth material beneath the level of the toe of the slope is unable to support the weight of the overlying material, the failure occurs along a surface passing beneath and beyond the toe; that type of slide is known as a base failure. Slides of both types can be stabilized with external restraint, but the position of the wall and accessory tiebacks is of prime importance. Obviously the two types call for walls that will resist forces acting in different directions. A wall used for restraining a slope failure is called upon to resist a force with a large horizontal component; conversely, a wall restraining a base failure must act more as a buttress or counterweight. In either case, building the wall to prevent failure on one predicted surface may result in a new failure surface passing above or below the wall, as shown in Figure 6A. Although the chance of the potential failure circle going under the wall as shown in Figure 6A can be negated by using a tieback on the wall, tiebacks are not very effective in preventing a failure developing above the wall.

Slumps vary in volume from only a few cubic yards for shallow surficial failures to many thousands of cubic yards for deep-seated failures. Major landslides commonly begin as relatively small slumps that grow in size by progressive failure. Slumps often degenerate into flows that extend well beyond the toe of the slope; such failures fit Varnes' (38) definition of a complex landslide. Retaining walls have been successful for containing relatively small slumps, but usually are not economical in developing enough resisting force in a wall to prevent or correct slumps that are very large.

Somewhat greater resistance to failure has been demonstrated by cylinder pile walls consisting of large-diameter reinforced concrete cylinders. Two well-known examples of their successful use to correct slides were to protect freeways in San Francisco and Seattle. At Potrero Hill in San Francisco, an anchored 800-ft- (244-m-) long and 20- to 60-ft- (6- to 18-m-) high retaining wall was constructed in 1966-67 to stabilize an Interstate Highway 280 cut (33). The grouted anchors for this wall were embedded a distance of 30-45 ft (9-14 m) into

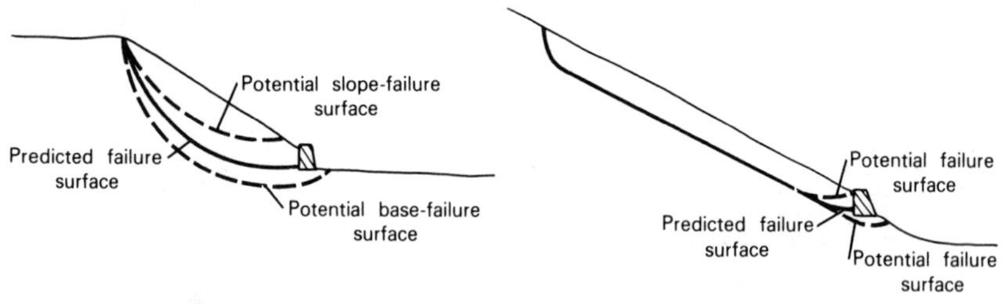

Figure 6.--Toe walls for stabilization of slides, showing possibility of development of failure surfaces above or below the predicted surface.

interbedded sandstone and shale retained by the wall. Within a few months parts of the wall failed because the failure surface passed beneath the anchors. In 1968, stability was restored to the cut by installation of 30 4-ft- (1.2-m-) diameter, 70- to 90-ft- (21- to 27-m-) long, cast-in-place concrete cylinders reinforced with 36 WF 230 steel beams. Similarly, cuts 40-50 ft (12-17 m) high for Interstate Highway 5 in Seattle in overconsolidated lacustrine clays overlain by glacial till were stabilized in the early 1960's by the use of large-diameter 3- to 13-ft (1- to 4-m) drilled, cast-in-place reinforced concrete cylinders spaced to form an almost continuous wall (22).

An early example of unsuccessful use of cylinder-pile walls was on the Portuguese Bend slide of the Palos Verdes Peninsula in Los Angeles County (20). The Portuguese Bend slide was a complex landslide involving both rotational and translational movement in the tuffaceous Miocene Altamira Shale Member of the Monterey Shale. Sliding had been reactivated in a 300-400 acre (120-160 ha) part of an ancient slide mass three times that large by grading in the upper part of the area. An attempt was made to stop the movement by installation of 25 4-ft- (1.2-m-) diameter reinforced-concrete caissons placed in vertical holes intersecting the failure surface. The exact manner of failure of the caissons is not known. Some tilted; some remained more or less fixed and were by-passed by landslide material in plastic flow; and others probably failed by crushing or shearing. Obviously the increase in shear strength provided by the caissons was very small compared to the stress applied to them by the large landslide mass moving along a previously existing failure surface. The landslide is continuing to move more than 20 years after the stabilization attempt.

Translational Slides

> "The distinction between rotational and translational slides is useful in planning control measures. The rotary movement of a slump, if the surface of rupture dips into the hill at the foot of the slide, tends to restore equilibrium in the unstable mass; the driving moment during movement decreases and the slide may stop moving. A translational slide, however, may progress indefinitely if the surface on which it rests is sufficiently inclined and as long as the shear resistance along this surface remains lower than the more or less constant driving force." (38, p. 14)

Translational slides occur in both rock and soil as movement approximately parallel to the surface of the slope. Failure takes place along joints, faults, bedding planes, contacts between different materials, or other geologic discontinuities. In most cases, the toe of the sliding mass slides out upon the original ground surface downslope of the failure surface.

Translational slides in rock fail along discontinuities such as shear zones, dipping joints, bedding planes, or combinations of these. Such slides may be actuated by the cutting of a slope by excavation or erosion; they occur when the inclination of the slope exceeds the angle of internal friction of the rock mass along the discontinuities (3). Slides on such slopes can also be triggered by excess pore pressures from water in joints or faults and along bedding surfaces.

In clay soils, translational failures take place along saturated sand or silt seams, particularly where these zones of weakness dip roughly parallel to the existing slope. Translational slides also are common along sloping contacts between soil and rock or at contacts between weathered and unweathered zones in soil.

Translational slides vary in size from small surficial soil slips to huge rock slides or rock block slides that are hundreds of millions of cubic yards in volume. Only small masses can be controlled by the resisting forces able to be developed by retaining walls, no matter what type of wall is used.

As shown in Figure 3, translational slides have a wide range of velocities from extremely slow to extremely rapid. The anticipated velocity of failure, which is a function of the material properties, slope angle, and length of runout, should be given consideration in selection and design of any restraining structure.

As mentioned earlier in the case for slumps, failure surfaces for translational slides have the discouraging tendency of changing path to go either under or over retaining walls intended to stabilize the slides (Figure 6B). In addition, translational slides, as do slumps, may degenerate to flows, if dilated slide debris becomes saturated.

Lateral Spreads

Lateral spreads occur in both rock and soil. The dominant mode of movement is lateral extension accommodated by shear or tensile fractures (38). The most common type of lateral spread is caused by liquefaction or plastic flow of underlying material. Where movement of a relatively coherent surface material occurs due to liquefaction of a thin subsurface layer, the spread can have the characteristic of a retrogressive translational slide.

Lateral spreads in bedrock are most usual where a thick layer of coherent rock overlies a weaker rock such as soft shale. Varnes (38) noted examples of spreading along escarpments and edges of plateaus in the USSR and Libya where an underlying shale or claystone became plastic and flowed to some extent, allowing the overlying firmer rock to separate into strips or blocks. Even though the rate of movement of such spreads in bedrock is extremely slow, structural restraint is not feasible because this type of landslide is ordinarily very large, perhaps measuring miles in extent.

Lateral spreads in soils commonly occur on shallow slopes. The initial failure may be a slump along a streambank or shoreline and progressive failure extends inland retrogressively, or failure may occur in silts or sensitive clays that lose strength during remolding. Failure by spreading also is common in varved clays where underlying or interbedded thin layers of sand may have practically no shear strength because of high pore-water pressures.

Another soil process leading to possible spreading is liquefaction of saturated sand due to dynamic stress imparted by earthquakes. This process is common in earthquake-prone areas such as Japan and western North America. Depending on the topography of the locality at which the liquefaction occurs, the process may or may not lead to lateral spreading. In general, liquefaction on unbounded flat surfaces results in foundation bearing-capacity problems, liquefaction on slopes with dips of a few degrees or on flat surfaces bounded by areas slightly lower will lead to lateral spreading, and liquefaction on steeper slopes or on flatter surfaces bounded by areas considerably lower can result in earth flow.

Spreads in either rock or soil generally occur on a scale too large to be retained by a wall. Predicting where and when a spread might occur is often difficult. However, in the case of soils such as sensitive clays or saturated sands and silts that are particularly susceptible to sudden reduction of strength and possible liquefaction, an understanding of the character of the soil can help considerably in determining whether or not walls built to retain the soil need to be designed to prevent lateral spreading. Youd and Perkins (43) provided insight into determining the potential for liquefaction of natural soils. In the case of fill materials, attention should be paid to poorly compacted sands and silts, and particularly to those placed hydraulically. Where there is a possibility of liquefaction and resulting lateral spreading, the soil should be drained, if possible; if the soil cannot be drained, any walls used to retain it should be designed to withstand full fluid pressures.

Because landslides reported in sensitive clays, such as those in Scandinavia and Canada, have generally been large, retaining walls commonly have been considered of little value in preventing or controlling them. However, as pointed out by Janbu (17), these large sensitive clay slides commonly start as insignificant movements that spread retrogressively in all directions. Seemingly, retaining walls might prove successful if placed in key positions to prevent the original small-scale movement. For potential sensitive-clay slides, this key position is usually at the toe of the susceptible slope, and particularly at the location of a new cut.

Case histories are available for failures of walls designed to retain poorly compacted, saturated granular fills when these fills liquefied during earthquakes. A well-documented example was the failure of gravity quay-walls and sheet-pile seawalls at Puerto Montt, Chile, as a result of the 1960 Chilean earthquake (7). Similar failures of quay-walls occurred in the 1964 Niigata, Japan, earthquake (11, 15).

Broms and Bjerke (4) noted the danger in using cylinder-pile walls to retain the soft soils commonly subject to spreading failure. They described failure of a 36-ft- (11-m-) high cast-in-place reinforced concrete pile wall built to retain a soft clay in central Sweden by the clay being squeezed like toothpaste between three adjacent piles.

Flows

Characteristically, the movement occurring in flow-type slope failures is much like that of a viscous liquid in that failure does not occur along any one shear surface or series of shear surfaces. As shown in Figure 3, these slope failures can range in velocity from extremely slow to extremely rapid. The slowest flow is commonly referred to as creep, and the fastest flow is similar in behavior to a free-flowing liquid. Flows occur in both rock and soil under both wet and dry conditions.

Flows in bedrock are characterized by deformations in either soft or hard rock that are distributed among many large or small fractures without concentration along a through-going fracture (38). Such movements generally are extremely slow and commonly are categorized as creep. Rock creep commonly occurs on a scale much too large to be retained by walls.

Flows in soil are more easily recognized as flows than are those in rock because the displacements within the mass are commonly larger than in rock and occur at greater velocity, and the appearance of the moving mass is much more that of a fluid (38). Slip surfaces within the moving mass generally are not discrete, and the basal boundary between the flowing mass and underlying material in place may be a sharply delineated surface or a zone of distributed shear.

Soil flows often take place during or after heavy rainfalls. The rate of movement of soil flows ranges from extremely slow to extremely rapid, depending on water content and degree of slope. In general, the most rapid soil flows are those with very high water contents, but totally dry granular soils on very steep slopes also can attain high flow velocities.

As shown in Figure 1, soil flows can be divided into debris flows and earth flows. A common type of earth flow is the so-called mudflow, a term often confused with debris flow. Debris flows are coarser than mudflows; the term mudflow should be reserved for an earth flow consisting of soil that is wet enough to flow rapidly and that contains at least 50 percent fines--i.e., sand-, silt-, and clay-size particles (38). In the United States debris flows are particularly common in the semiarid southwestern states where dry soils on steep slopes are easily activated by any heavy rainfall. Southern California is particularly hard hit by catastrophic soil flows; for example, during the period 1962-71, 23 people in the greater Los Angeles area died by being buried or struck by debris flows resulting from heavy rainstorms (5).

As noted above, flows often develop as part of a complex landslide process beginning as falls, topples, slides, or spreads. This process is particularly common in the slumping or sliding failure of very wet soils that degenerate into flows.

Traditional retaining walls are effective for prevention or control of small, creeping masses of rock or soil. For the more-fluid, rapidly moving flow failures, however, restraining structures have not generally been considered to be effective (2). However, check dams, which are a type of retaining wall that can control highly fluid earth flows by providing "reservoir" volume, have been used commonly in Japan, Indonesia, the U.S.S.R., and other countries to control debris flows caused by volcanic eruptions or heavy rainfall. Check dams also have been used to control debris flows in southern California (8). These dams are designed to sustain a "full reservoir" fluid pressure plus dynamic loading due to arrival of the flow. Aoki (1) provided data for design of a check dam for restraint of a debris flow.

Even though traditional retaining walls cannot completely control rapidly moving soil flows, deflection walls can be used to protect property or buildings in the path of potential flows. A deflection wall is a retaining wall that is placed at an angle other than 90° to predicted direction of flow (14). The purpose of such positioning is to decrease the perpendicular force component on the wall and to deflect the flow around the property to be protected. Figure 7 illustrates a typical layout for a deflection wall.

Even with the use of a deflection wall, further protection of houses and other buildings from flows may be advisable by designing the uphill walls of the buildings as retaining walls. Suwa et al. (34), Mears (19), and Hollingsworth and Kovacs (14) provide design pressure data for such walls.

Mears (19) also noted that the impact forces from debris flows can be made relatively harmless if the large boulders can be separated from the flows. In some cases, this separating function can be performed with "slotted" steel and reinforced concrete retaining walls in which vertical members are spaced so as to allow passage of smaller boulders while retaining the larger ones.

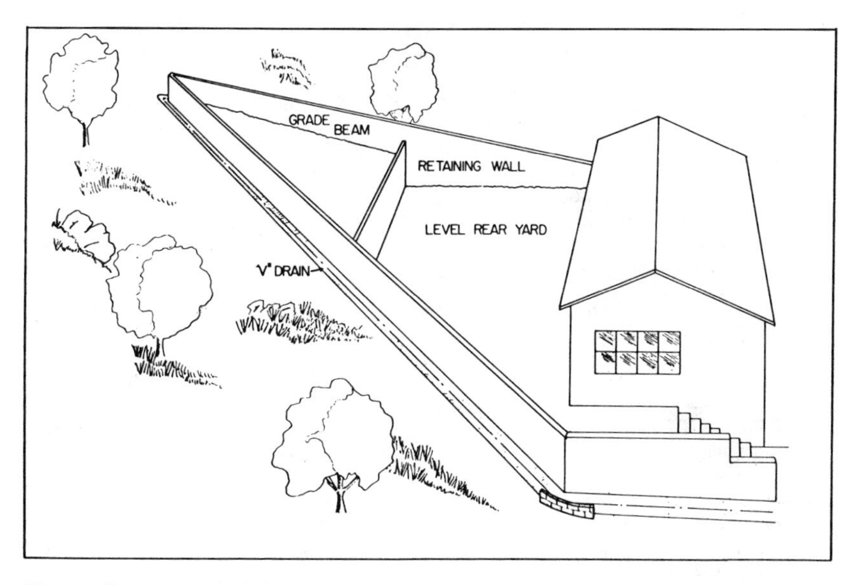

Figure 7.--Typical deflection wall layout for protection of house and yard from possible debris flow (after reference 14).

CONCLUSIONS

In general, the selection of remedial measures for any type of potential or existing landslide requires a thorough understanding of the expected landslide process, which, in turn, demands a knowledge of the local geology that determines the properties and distribution of the earth materials involved.

Retaining walls of various types have been used, with mixed success, to prevent or to control rotational and translational slides. They have not seen extensive use as remedial measures for falls, topples, spreads, or flows; more certain anticipation and better understanding of the characteristics of these types of landslides may lead to increased application of walls to their prevention and control.

Appendix--References

1. Aoki, K., "Eruption of Mt. Usu and Erosion Control Works," presented at the 1981, 2nd Japan-Indonesia Symposium on Volcanic Debris Flow Treatment, held at Djakarta, Indonesia, 33 pp.
2. Baker, R. F., and Marshall, H. E., "Control and Correction," Landslides and Engineering Practice, E. B. Eckel, ed., Special Report 29, Highway Research Board, National Academy of Sciences-National Research Council, Washington, D.C., 1958, pp. 150-188.
3. Broms, B. B., "Landslides," Foundation Engineering Handbook, H. F. Winterkorn, and H. Y. Fang, eds., Van Nostrand Reinhold Company, New York, N.Y., 1975, pp. 373-401.
4. Broms, B. B., and Bjerke, H., "Extrusion of Soft Clay Through a Retaining Wall," Canadian Geotechnical Journal, Vol. 10, No. 1, February, 1973, pp. 103-109.
5. Campbell, R. H., "Soil Slips, Debris Flows, and Rainstorms in the Santa Monica Mountains and Vicinity, Southern California," Professional Paper 851, U.S Geological Survey, 1975, 51 pp.
6. de Freitas, M. H., and Watters, R. J., "Some Field Examples of Toppling Failure," Geotechnique, Vol. 23, No. 4, 1973, pp. 495-514.
7. Duke, C. M., and Leeds, D. J., "Response of Soils, Foundations, and Earth Structures to the Chilean Earthquakes of 1960," Bulletin of the Seismological Society of America, Vol. 53, No. 2, February, 1963, pp. 309-357.
8. Farrell, W. R., "Report on Debris Reduction Studies for Mountain Watersheds of Los Angeles County," Los Angeles County Flood Control District, November, 1959, 164 pp.
9. Gould, J. P., "Lateral Pressures on Rigid Permanent Structures," State-of-the-Art Papers, presented at the 1970, ASCE Lateral Stresses in the Ground and the Design of Earth-Retaining Structures Specialty Conference, held at Cornell University, Ithaca, N.Y., pp. 219-269.
10. Groupe d'Etudes des Falaises, "Eboulements et Chutes de Pierres sur les Routes--Recensement des Parades," Rapport de recherche LPC No. 81, Laboratoire Central de Ponts et Chaussees, Paris, France, July, 1978, 252 pp.
11. Hakuno, M., "Harbor facilities," in General Report on the Niigata Earthquake, H. Kawasumi, ed., Tokyo Electrical Engineering College Press, Tokyo, 1968, pp. 483-497.

12. Hansen, W. R., "Effects of the Earthquake of March 27, 1964, at Anchorage, Alaska," Professional Paper 542-A, U.S. Geological Survey, 1965, 68 pp.
13. Henkel, D. J., "Local Geology and the Stability of Natural Slopes," Journal of the Soil Mechanics and Foundations Division, ASCE, Vol. 93, No. SM4, July, 1967, pp. 437-446.
14. Hollingsworth, R., and Kovacs, G. S., "Soil Slumps and Debris Flows: Prediction and Protection," Bulletin of the Association of Engineering Geologists, Vol. 18, No. 1, 1981, pp. 17-28.
15. Ichihara, M., "River structures," General Report on the Niigata Earthquake, H. Kawasumi, ed., Tokyo Electrical Engineering College Press, Tokyo, 1968, pp. 503-515.
16. Jahns, R. H., "Landslides," Geophysical Predictions," National Academy of Sciences, Washington, D.C., 1978, p. 58-65.
17. Janbu, N., "Critical Evaluation of the Approaches to Stability Analysis of Landslides and Other Mass Movements," Proceedings, International Symposium on Landslides, New Delhi, India, Vol. 2, 1980, pp. 109-128.
18. Ladd, G. E., "Methods of Controlling Highway Landslides," Roads and Streets, Vol. 68, No. 11, 1928, pp. 529-538.
19. Mears, A. I., "Debris-flow Hazard Analysis and Mitigation--An Example from Glenwood Springs, Colorado," Information Series 8, Colorado Geological Survey, Denver, Colorado, 1977, 45 pp.
20. Merriam, R., "Portuguese Bend Landslide, Palos Verdes Hills, California," Journal of Geology, Vol. 68, 1960, pp. 140-153.
21. Natarajan, T. K., Bhandari, R. K., Rao, E. S., and Singh, A., "A Major Landslide in Sikkim--Analyses, Correction and Efficacy of Protective Measures," Proceedings, International Symposium on Landslides, New Delhi, India, Vol. 1, 1980, pp. 397-402.
22. Palladino, D. J., and Peck, R. B., "Slope Failures in an Over-consolidated Clay, Seattle, Washington," Geotechnique, Vol. 22, No. 4, 1972, pp. 563-595.
23. Peckover, F. L., and Kerr, J. W. G., "Treatment and Maintenance of Rock Slopes on Transportation Routes," Canadian Geotechnical Journal, Vol. 14, No. 4, November, 1977, pp. 487-508.
24. Piteau, D. R., and Peckover, F. L., "Engineering of Rock Slopes," Landslides--Analysis and Control, R. L. Schuster, and R. J. Krizek, eds., Special Report 176, Transportation Research Board, National Academy of Sciences, Washington, D.C. 1978, pp. 192-228.
25. Ritchie, A. M., "Evaluation of Rockfall and Its Control," Highway Research Record, Highway Research Board, National Academy of Sciences-National Research Council, No. 17, 1963, pp. 13-28.
26. Royster, D. L., "Landslide Remedial Measures," Bulletin of the Association of Engineering Geologists, Vol. 16, No. 2, 1979, pp. 301-352.
27. Schuster, R. L., "Introduction," Landslides--Analysis and Control, R. L. Schuster and R. J. Krizek, eds., Special Report 176, Transportation Research Board, National Academy of Sciences, Washington, D.C., 1978, pp. 1-10.
28. Schweizer, R. J., and Wright, S. G., "A Survey and Evaluation of Remedial Measures for Earth Slope Stabilization," Research Report 161-2F, Center for Highway Research, The University of Texas at Austin, August, 1974, 123 pp.

29. Shearer, C. F., Taylor, F. A., and Fleming, R. W., "Map Showing Distribution and Costs of Landslides in San Diego County, California, During the Winters of 1978-79 and 1979-80," Miscellaneous Field Studies Map, U.S. Geological Survey, in preparation.
30. Skempton, A. W., "The Rate of Softening in Stiff, Fissured Clays, with Special Reference to London Clay," Proceedings of the Second International Conference on Soil Mechanics and Foundation Engineering, Rotterdam, Vol. 2, 1948, pp. 50-58.
31. Skempton, A. W., and Hutchinson, J., "Stability of Natural Slopes and Embankment Foundations," State of the Art Volume, Seventh International Conference on Soil Mechanics and Foundation Engineering, Mexico City, 1969, pp. 291-340.
32. Skempton, A. W., and La Rochelle, P., "The Broadwell Slip: A Short-term Failure in London Clay," Geotechnique, Vol. 15, No. 3, 1965, pp. 221-242.
33. Smith, T., and Forsyth, R., "Potrero Hill Slide and Correction," Journal of the Soil Mechanics and Foundations Division, ASCE, Vol. 97, No. SM3, March, 1971, pp. 541-564.
34. Suwa, H., Okuda, S., and Yokoyama, K., "Observation System on Rocky Mudflow," Bulletin of the Disaster Prevention Research Institute, Kyoto University, Kyoto, Japan, Vol. 23, Parts 3-4, No. 213, December, 1973, pp. 59-73.
35. Terzaghi, K., "Mechanism of Landslides," Application of Geology to Engineering Practice, Berkey Volume, Geological Society of America, 1950, pp. 83-123.
36. Terzaghi, K., and Peck, R. B., "Soil Mechanics in Engineering Practice," 2nd ed., John Wiley & Sons, Inc., New York, N.Y., 1967, 729 pp.
37. Varnes, D. J., "Landslide Types and Processes," Landslides and Engineering Practice, E. B. Eckel, ed., Special Report 29, Highway Research Board, National Academy of Sciences-National Research Council, Washington, D.C., 1958, pp. 20-47.
38. Varnes, D. J., "Slope Movement Types and Processes," Landslides--Analysis and Control, R. L. Schuster and R. J. Krizek, eds., Special Report 176, Transportation Research Board, National Academy of Sciences, Washington, D.C., 1978, pp. 11-33.
39. Varnes, D. J., and The Commission on Landslides and Other Mass Movements on Slopes, International Association of Engineering Geology, "Landslide Hazard Zonation--A Review of Principles and Practice," UNESCO, Paris, in press.
40. Veder, Christian, Landslides and their Stabilization, Springer-Verlag, New York, N.Y., 1981, 247 pp.
41. Voight, Barry, "Rockslides and Avalanches," Vol. 1, Natural Phenomena, Elsevier, New York, N.Y., 1978, 833 pp.
42. Ward, W. H., "The Stability of Natural Slopes," The Geographical Journal, The Royal Geographical Society, London, Vol. 105, Nos. 5, 6, May-June, 1945, pp. 170-197.
43. Youd, T. L., and Perkins, D. M., "Mapping Liquefaction-Induced Ground Failure Potential," Journal of the Geotechnical Engineering Division, ASCE, Vol. 104, No. GT4, April, 1978, pp. 433-446.

Note: *Figures 1, 2A, 2C, 2F, 2H and 4 have been reproduced by courtesy of the Transportation Research Board, Washington, D.C.*

THE ANALYSIS OF WALL SUPPORTS TO STABILIZE SLOPES

by

Norbert R. Morgenstern,[1] M.ASCE

INTRODUCTION

Retaining structures are used increasingly to control slope stability problems. This paper is intended to present an overview of the main analytical considerations that enter into their design. Hutchinson (13) provides a starting point for the discussion.

He observes that the use of rigid restraining structures is generally less appropriate for stabilizing slopes than methods involving drainage or reshaping the slope. Numerous failures of such structures have been reported (2, 15) but it should be noted that most of these cases are old and that both our understanding of slope instability processes and our capability for constructing more effective walls have improved in recent years. Hutchinson observes that when properly engineered, rigid restraining structures can have a useful role, particularly where space is restricted and he indicates the following types have been used successfully:

a) Retaining walls founded beneath the unstable ground: examples of the variety of designs that may be used are found in (2 and 15).
b) Piles: a wall of continuous or closely spaced driven cantilever piles can be effective in stabilizing shallow slides (18) while anchored sheet or bored pile walls have been used successfully to stabilize deep-seated slides (1, 9).
c) Soil and rock anchors, generally pre-stressed: these are employed either in conjunction with retaining structures or alone to reduce the driving forces of a landslide and to increase the normal effective stresses on its slip surface (7).

In the following, it will be convenient to separate wall supports into three groups:

i) Free Standing Walls,
ii) Cantilevered Walls,
iii) Tied-back Walls,

and to discuss the analysis of each group separately. Algebraic details of the analysis will be omitted since they are readily accessible elsewhere in the literature.

---

[1] Professor of Civil Engineering, University of Alberta, Edmonton, Canada, T6G 2G7

FREE STANDING WALLS

Free standing walls are conventional retaining walls that gain their support by the action of gravity alone. They are passive elements in that they are subjected to lateral loading by the tendency for the adjacent ground to move. Principles for the analysis of free standing walls are well-known and need not be repeated here.

The height of gravity retaining walls that can be used to stabilize slopes is limited, particularly when seated on argillaceous material. When used to stabilize actual slides, only the residual angle of shearing resistance can be mobilized and the thrust exerted by the slide mass may be substantial. A tendency to underestimate this and a tendency to overestimate the sliding resistance of some walls has resulted in poor performance in the past. A height of 10m is about the upper limit for conventional gravity retaining walls and slides of only modest proportions can be prevented or stabilized using this type of restraining structure. Special attention should be given to the foundation conditions beneath the wall. In some circumstances, sliding resistance might have to be augmented by the provision of a shear key. Careful detailing of drainage to minimize water pressures acting on the wall is always required. Free standing walls are inappropriate for larger slides, slides which penetrate beneath the level accessible by excavation, and for circumstances where more positive loading of the soil behind the wall is desirable.

CANTILEVERED WALLS

Actual or potential slides can be stabilized by means of piles driven or bored into stable underlying soil. The construction of a stabilizing wall in this manner is not new, but it is only recently that it has been subjected to comprehensive analysis. At one extreme, Farris (10) describes the use of timber piles to stabilize railway embankment movements. Because of the low flexural strength of wood, they can only be considered for stabilizing small slides. At the other extreme, large diameter steel pipe piles infilled with very high strength concrete have been used in Japan (11). Steel wire cables are used for reinforcement to provide high flexibility. Continuous sheet piles and cast-in-place reinforced concrete bored piles have also been used.

Cantilevered walls are attractive for several reasons:

i) They may be constructed with a minimum of excavation; this is particularly advantageous in comparison with conventional retaining walls for cuts in urban areas where rights-of-way are restricted.

ii) When cast-in-place, the wall has the advantage of being installed without significantly decreasing slope stability during its construction. This is particularly beneficial in strain-softening soils.

The major disadvantage associated with the use of cantilevered walls to stabilize slopes is cost. Generally, drainage or reshaping will prove more attractive. There are usually extenuating circumstances that make a cantilevered wall the appropriate solution to stabilize a slope.

Unsuccessful slope stabilization using a cantilevered wall may be due to any of the following:

i) failure below the wall;
ii) failure of pile or wall by bending, tilting or shearing;
iii) failure as a result of shearing soil flow between piles;
iv) failure from induced pore pressures.

The first two modes of failure are eliminated by adequate design to satisfy both slope stability and pile stability. Soil flow may be prevented by proper spacing in a pile system. Failure from induced pore pressures is generally restricted to cases where piles are driven into soft or sensitive clay slopes. Reductions in available shear strength of some 20 - 30% have been recorded in cases where driving was uncontrolled (4, 6).

The use of cantilevered walls to stabilize slopes is encouraged by several successful case histories. Only a few will be noted here.

Serious movement resulted from excavations into the heavily overconsolidated and fractured lacustrine silty clay of the Seattle region. To prevent these movements, the Washington State Highway Department installed, before further excavation, a cylinder pile wall placed uphill from the proposed cut. No analysis of the increase in Factor of Safety of the slope due to the presence of the wall appears in the literature. An analysis of the pile design has been given (1) and subsequent monitoring attests to the satisfactory performance of the restraining wall (20). The pile spacing was small enough to preclude soil flow.

DeBeer and Wallays (9) have also utilized contilevered piles to stabilize a moving slope in weathered shales. The basis for calculation has been given, and it is presumed that the performance has been satisfactory.

The case history presented by Sommer (17) is particularly valuable. In this instance, movements beneath a highway embankment were stabilized by the installation of piers 3m in diameter, installed 5m below the slip surface and spaced at 9m intervals. This resulted in only a modest increase in overall Factor of Safety, but it was adequate to stop the movements. It is more difficult to evaluate the load carrying requirements of piles installed to augment the stability of an already stable slope. In these circumstances an understanding of the deformation process leading to failure is particularly valuable and finite element calculations are useful to explore the build-up of loads on the wall as deformations take place.

In order to undertake the analysis of a cantilevered wall used to stabilize a slope, it is necessary to conduct analyses of both slope and pile stability. This has recently been reviewed by Viggiani (19) in a valuable study who indicates the following steps:

i) Evaluate the total shear force needed to increase the Factor of Safety of the slope by the desired amount.
ii) Evaluate the maximum shear force that each pile can receive from the sliding soil mass and transmit to the stable underlying soil.
iii) Select the type and number of piles and their most suitable location in the slope.

The required shear force is readily estimated by means of conventional limit equilibrium analysis. Viggiani (19) provides guidance on the analysis of pile stability. For simplicity, he considers a horizontal layer of soil of thickness ℓ_1 sliding on a slip surface above a strong underlying soil. Both soils are assumed to be saturated clays with undrained shear strength C_1 above the slip surface and C_2 below the slip surface. As shown in Figure 1, a pile crosses the sliding soil and the slip surface, extending to a depth $\ell_2 = \lambda\ell_1$ beneath it. Following arguments originally extended by Broms (5), when yielding occurs

$$P_y = \kappa\ c\ d \qquad (1)$$

where P_y denotes the yield value of the pile-soil interaction
d denotes the pile diameter
κ denotes a bearing capacity factor

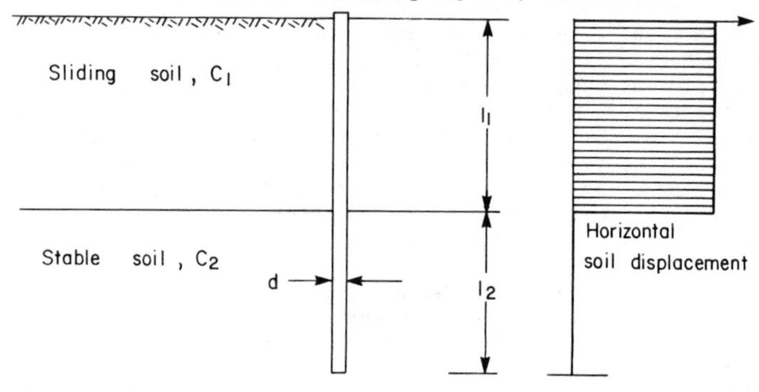

Figure 1 Yielding Around Pile

DeBeer (8) has reviewed values for κ and shown that they depend upon whether the pile is displaced in a non-moving soil or whether the pile is in flowing soil. Although there is considerable variance between existing theories, values of κ above and below the slip surface appear to be about 4 and 8 respectively.

In a systematic study of failure modes, Viggiani notes that if the pile is rigid, failure can occur by one of three modes. As illustrated in Figure 2, displacements can occur so that the yield value develops only below the slip surface. Another mode involves rotation with yielding both above and below the slip surface. Finally, if the pile were fixed in firm soil, the upper material could flow around it. In each case the soil reactions differ and result in different shear force and bending movement diagrams for the pile. However in most practical problems, the maximum moment to be taken by the pile constitutes a limiting factor and other failure modes, including one or two plastic hinges, can be envisaged as shown in Figure 2.

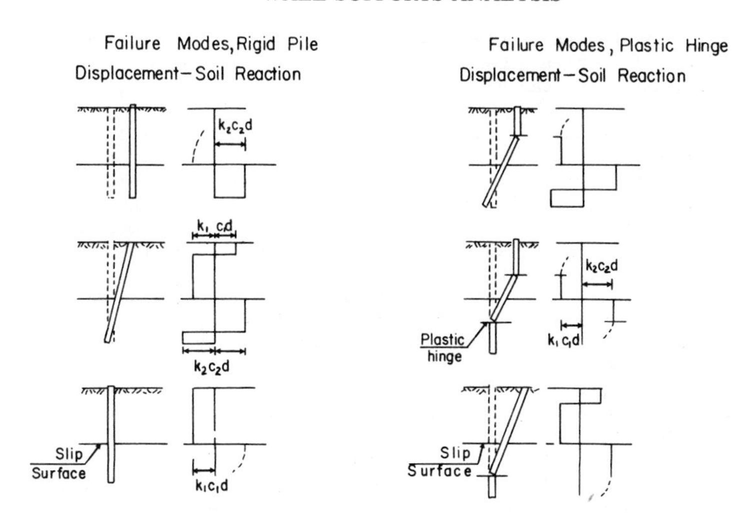

Figure 2 Ultimate Loads on Piles in Landslides. after Viggiami (19)

Of particular interest in this work is the comparison made between the implications of these various modes with case histories. In all of the case histories analysed, the yield moment of the pile was the limiting factor and in three cases out of four, a plastic hinge formed below the slip surface. Values of κ deduced from the case histories were in reasonable agreement with those noted above.

Ito, Matsui and Hong (14) indicate that factors such as the interval between piles, the fixity condition of the pile head, the pile length above the sliding surface, the pile diameter and the stiffness of the pile all have an influence on the action of piles used to stabilize landslides and they provide guidance on how these factors can be considered in design in a systematic manner.

TIED-BACK WALLS

Tied-back walls to stabilize slides may be driven or cast-in-place. Generally some flexibility in the wall system is desirable if significant movements are anticipated. Tied-back walls have a major advantage over systems discussed previously in that the restraining system actively opposes the movement of the soil mass rather than behaving in a passive manner. As a result a component of the thrust of the tie-backs acts against the thrust of the potential sliding mass while another component of the thrust increases the normal stresses on slip surfaces in the soil. Both actions deter downslope movements. There are also less readily quantifiable benefits associated with the use of tie-backs. By increasing the compression in the soil, softening processes that tend to weaken the soil with time are inhibited and the soil mass is forced to act in a more integral manner.

Current practice in the design of tied-back walls to stabilize slopes is based on either a consideration of Factor of Safety alone or on a combination of deformation and limit equilibrium considerations. The case history of stabilizing a slide in Switzerland presented by Huder and Duerst (12) is an example of the former procedure while the design of the tied-back wall for the Edmonton Convention Centre (3) illustrates the latter.

Figure 3 indicates the circumstances of the slide that threatened a section of the Swiss Federal Railways. The slope deposits were heterogeneous talus with a matrix comprised of clay having a Plasticity Index of 18.5% and a residual angle of internal friction of 23.5°. The wall system chosen was an intermittent bored pile wall using permeable concrete as lagging to facilitate drainage behind the wall. The spacing between the 1.16m reinforced concrete piles was 2.50m. The piles were about 13m long and were embedded at least 3m in the bedrock. As shown in the Figure, multiple levels of tie-backs were needed to achieve the design Factor of Safety.

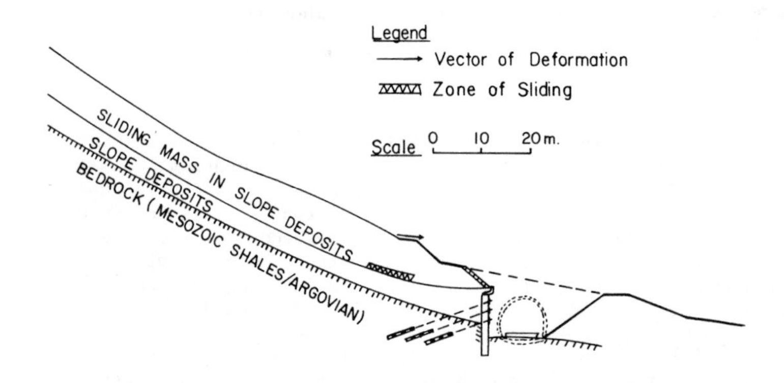

Figure 3 Stabilization of Swiss Railway Slide after Huder and Duerst (12)

The tie-back wall was designed for a Factor of Safety of 1.5 using conventional limit equilibrium considerations while the anchors themselves possessed a Factor of Safety of 2.0 against failure, in accordance with the relevant Swiss code (16). The anchors were inclined at 20° to the horizontal and were fixed in competent rock. Since the anchors were permanent, provision of corrosion protection was necessary and various tests and monitoring of anchor behaviour were undertaken as required by the code.

To this extent the case history is typical of the design of many tied-back walls. However during initial excavations for anchor tests, movements were aggravated and it became possible to determine accurately the slip mechanism. Inclinometers revealed a slip surface at a depth of 8 to 10m within the soil mass, and not at the soil-rock interface. As suggested in Figure 3, the restraining effect of the wall actually contributes very little to increasing the stability of the slope.

Substantial movements developed during the installation of the wall but they have decreased since completion. Original studies showed that although the talus was unsaturated, in general, water could be found entrapped in various disconnected layers. The decrease in movement following construction of the wall has been attributed to a reduction of pore water pressures in layers close to the sliding zone. It has transpired that the wall is a stabilizing factor more for its role in enhancing drainage than for the provision of positive anchor forces. Since high pore water pressures could still trigger large movements, a continual monitoring program has been implemented to eliminate danger due to potential movements.

In addition to ensuring that the tied-back wall provides an adequate overall Factor of Safety against potential slip mechanisms, circumstances arise when the deformations at the top of the wall also have a bearing on the analysis. One example where such considerations may govern occurs when major structures exist adjacent to the wall and deformations must be limited to avoid damaging them. In the case of the Edmonton Convention Centre (3) deformations of the wall were an important factor in the analysis, but not due particularly to the adjacent structures. The excavation was being made in a hillside notorious for its past record of instability. Regional studies had indicated that long-term softening was an important process leading to instability. In order to inhibit this process in the slopes adjacent to the excavation, it was judged desirable to null the movements of the ground near the wall insofar as possible. This led to the earth pressure loading diagram shown in Figure 4 and the configuration of tie-backs consistent with it.

As indicated in Figure 4, various potential slip mechanisms could be envisaged and limit equilibrium analyses were conducted to ensure that the Factor of Safety was high enough. Values in the range 1.75 to 2.0 were being sought. The lower most slip mechanism, governed by the presence of a bentonite layer, had a Factor of Safety of about 1.8.

In addition to limit equilibrium analyses, finite element calculations were undertaken. These calculations took into account the pre-existing stress field, the detailed stratigraphy, the actual sequence of excavation and tie-back installation, and the presence of horizontal layers weak in shear. In an iterative manner, it was shown that good ground control should be possible with the design that was ultimately adopted and that proved to be the case. The wall has performed in an excellent manner.

However local yielding in the lower most bentonite layer developed to a degree greater than anticipated. Special attention was given to this layer in both the site investigation and the analysis that led to the view that the bentonite at depth behind the wall was not sheared in-situ and would not pass its peak resistance in response to

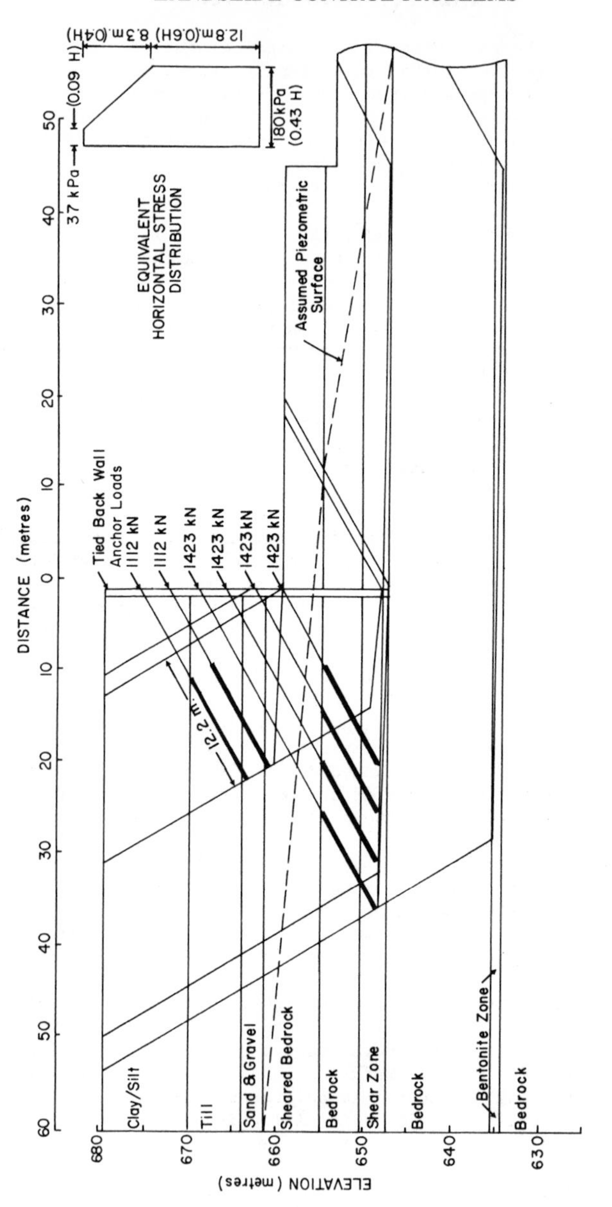

FIGURE 4 OVERALL STABILITY ANALYSIS, EDMONTON CONVENTION CENTRE

shear induced by the excavation. This has proved not to be the case and it has been a salutory lesson in exposing our analytical limitations when dealing with problems of progressive failure. When progressive failure develops in response to excavation, the available shearing resistance following excavation to install a retaining wall may become less than determined at the outset and the overall Factor of Safety is reduced accordingly. In the case of the Edmonton Convention Centre, an initial overall Factor of Safety of about 1.8 was insufficient to inhibit this process even at a depth many metres below the base of the excavation.

Geotechnical considerations also enter into the design of the wall itself, whether it is a tangent pile wall (3), an intermittent pile wall (12) or some other configuration. Analysis should ensure that the wall is stable in the vertical direction. If the wall is modelled as part of a finite element analysis, the stresses calculated provide direct input into the structural analysis of the wall. Otherwise, reasonable results with respect to bending and shear stresses in the wall can usually be obtained by analyzing it as a beam on an elastic foundation.

CONCLUDING REMARKS

1) Free standing walls are limited in their value as restraining structures to stabilize slopes.
2) There has been progress both in understanding the action of cantilevered walls within a slope and in gaining experience with them. Analysis requires separate consideration of both slope stability and pile stability.
3) Where practical, tied-back walls have advantages over alternative wall support systems. There is increasing confidence in relying on high capacity permanent anchors in both soils and rocks.
4) In every instance, it is essential that the specifics of the slope instability mechanism be understood prior to selecting a wall support system and designing it in detail.
5) Many wall support systems require excavation. If the circumstances are such that progressive failure is possbile, it should be recognized that routine analyses for progressive failure are not at hand. This is a serious limitation to the rational analysis of wall supports used to stabilize slopes and must be compensated for in the overall design and monitoring program.

APPENDIX I - REFERENCES

1. Andrews, G.H. and Klassell, J.A., "Cylinder Pile Retaining Wall," Highway Research Record No. 56, 1964, pp. 83-97.
2. Baker, R.F. and Marshall, H.C., "Control and Correction, In: Landslides and Engineering Practice," ed. by E.B. Eckel, Highway Research Board Special Report No. 29, 1958, pp. 150-188.
3. Balanko, L.A., Morgenstern, N.R. and Yachyshyn, R., "Tangent Pile Wall for Edmonton Convention Centre," Proceedings this Symposium, ASCE, 1982.
4. Brand, E.W. and Krasaesin, P., "Investigation of an Embankment Failure in Soft Clay," Geotechnical Engineering, 1971, Vol. 2, pp. 53-66.
5. Broms, B.B., "Lateral Resistance of Piles in Cohesive Soils," Jnl. Soil Mechs. Found. Div., ASCE, 1964, Vol. 90, SM2, pp. 27-63.
6. Broms, B.B. and Bennermark, H., Discussion in Proc. Geotechnical Conf., Oslo, 1968, Vol. 2, pp. 118-120.
7. D'Appolonia, E., Alpenstein, R. and D'Appolonia, D.J., "Behavior of a Colluvial Slope," J.Soil Mechs. Found. Div., ASCE, Vol. 93, 1967, SM4, pp. 447-473.
8. DeBeer, E.E., "State-of-the-Art Report: Piles Subjected to Static Lateral Loading," in Proc. Specialty Session 10, 9th Int. Conf. Soil Mechs. Found. Eng., 1977, pp. 1-14.
9. DeBeer, E.E. and Wallays, M., "Stabilization of a Slope in Schists by Means of Bored Piles Reinforced with Steel Beams," Proc. 2nd Int. Cong. Rock Mechanics, Belgrade, 1970, Vol. 3, pp. 361-369.
10. Farris, J.E., "Roadbed Stabilization," Amer. Rail. Eng. Assoc., Bull. 76, 1975, pp. 419-431.
11. Fukuoka, M., "The Effects of Horizontal Loads on Piles Due to Landslides," in Proc. Specialty Session 10, 9th Int. Conf. Soil Mechs. Found. Eng., 1977, pp. 27-42.
12. Huder, J. and Duerst, R., "Safety Considerations for Cut in Unstable Slope," Proc. 10th Int. Conf. Soil Mechs. Found. Eng., Stockhom, 1981, Vol. 3, pp. 431-436.
13. Hutchinson, J.H., "Assessment of the Effectiveness of Corrective Measures in Relation to Geological Conditions and Types of Slope Movement," Bulletin of the International Association of Engineering Geology, No. 16, 1977, pp. 131-155.
14. Ito, T., Matsui, T. and Hong, W.P., "Design Method for Stabilizing Piles Against Landslide - One Row of Piles," Soils and Foundations, 1981, Vol. 21, pp. 21-37.
15. Root, A.W., "Prevention of Landslides, In: Landslides and Engineering Practice," ed. by E.B. Eckel, Highway Research Board Special Report No. 29, 1958, pp. 113-149.
16. SIA, Swiss Association of Engineers and Architects, "Boden und Felsanker," Norm 191, 1977, 40 p, Zurich.
17. Sommer, H., "Creeping Slope in a Stiff Clay," in Proc. Specialty Session 10, 9th Int. Conf. Soil Mechs. Found. Eng., 1977, pp. 113-118.
18. Toms, A.H. and Bartlett, D.L., "Applications of Soil Mechanics in the Design of Stabilizing Works for Embankments, Cuttings and Track Formations," Proc. Inst. Civ. Engrs., Vol. 21, 1962, pp. 705-732.

19. Viggiani, C., "Ultimate Lateral Load on Piles Used to Stabilize Landslides," Proc. 10th Int. Conf. Soil Mechs. Found. Eng., Stockhom, 1981, Vol. 3, pp. 555-560.
20. Wilson, S.D. and Mikkelsen, P.E., "Field Instrumentation" in Landslides: Analysis and Control, Transportation Research Board, Special Report 176, 1978, pp. 112-138.

Evaluation of Landslide Properties

by

Dwight A. Sangrey[1] M. ASCE

INTRODUCTION

Slope instability or landsliding occurs when a combination of important factors produces a situation where forces tending to move a mass of soil or rock downhill overcome forces and soil strength resisting this movement. A wall used to control landsliding is designed to produce a resisting force. The analysis or design of a wall used for slope stability depends upon understanding the other factors in the overall process of mass movement on slopes. These other factors can be summarized in three groups: a) the topography and stratigraphy of the site, b) shearing resistance of the materials and c) the level and forces associated with groundwater. Other factors, such as external loads or dynamic earthquake loading, may be important in some specific instances but will not be discussed in this paper.

Methods have been developed for evaluating the characteristics of the three factors noted above. Topography and stratigraphy are evaluated from surface and subsurface site surveys. Shearing resistance is determined from field tests or from laboratory tests on specimens taken from the site. The characteristics of groundwater are usually estimated from water levels measured in one or more wells drilled in the slide area. The current state-of-practice for evaluating these characteristics is well developed and is based on an extensive amount of research and documented experience in the field (17).

The objective of this paper is to provide an overview of methods for evaluating landslide properties, particularly shearing resistance and groundwater levels. Because of the extensive literature which can be referenced, emphasis is placed on the fundamental aspects of evaluating landslide properties. There are some important exceptions, but in most cases the addition of a wall to a slope stability problem will not change the value of, or method for evaluating, a landslide characteristic. However, the accuracy of a wall design or analysis will depend upon the uncertainty associated with the measurement of the landslide properties, so emphasis in this paper will be directed to the effects of walls.

---

[1] Professor and Head, Department of Civil Engineering, Carnegie-Mellon University, Schenley Park, Pittsburgh, PA 15213

STRENGTH PROPERTIES

For the limit equilibrium methods used in design and analysis of landsliding controlled by walls, soil strength properties must be assigned to all surfaces along which movement might occur. These include the slide surface in the soil or rock, the contact between the wall and earth materials, and the area around an anchor. There are some important distinctions concerning the development of shearing resistance along these different surfaces, which will be discussed later, but the underlying principles of soil behavior are the same. With few practical exceptions the shearing resistance of soils and rocks (τ) is modeled using the Mohr-Coulomb failure criterion (defined using effective stress):

$$\tau = c' + \sigma' \tan \phi'$$

where

σ' = effective normal stress on the failure surface

c' = cohesion intercept

ϕ' = angle of internal friction

An extensive scientific literature has developed on the subject of shear strength of soils and rocks with specific reference to landsliding (17-Chap. 6). Rather than reviewing this literature, it is sufficient to highlight the major principles applicable to control of landsliding by walls. These principles are essentially the same as apply to any landslide problem. Shear strength properties must be defined with great care giving particular attention to the changes in shearing resistance which occur with time, movement and drainage. Understanding and modeling these changes in a fundamental and rational way is the key to successful engineering control of landsliding.

Shear strength of soils and rocks can be described using either effective stress or total stress. Although there are some few situations where total stress methods might be satisfactory, the influence of time and drainage effects on slope stability are so important that effective stress methods must be used.

Design and analysis of slopes should consider the state of stress, and changes in the state of stress, which occur during the development of the slope. Various classes of soil behavior, such as progressive failure (3), are strongly influenced by changes in the state of stress. Although the specific levels of stress existing in the field are rarely measured, estimated in situ stress states are used along with more precise evaluations of stress change. One important application of this information is the design of laboratory and field testing programs to evaluate the shear strength properties.

When walls are used to control slope failures, the significance of stress state and stress change is increased. As an example, cutting a slope in a soil deposit will produce a reduction in stress and a rotation of principal stress axes (Fig. 1a). Both of these changes can significantly affect the strength properties of soils (1,14) and the

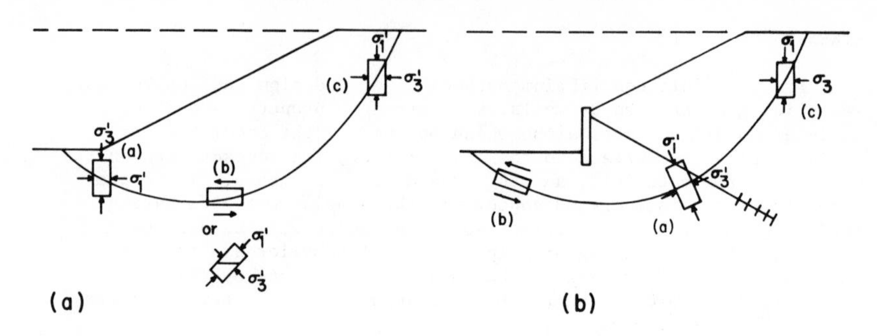

Figure 1. The distribution of stress along potential failure surfaces on a slope, a) for a typical excavation, b) for a wall with anchors.

appropriate methods for testing to evaluate these properties (17, Chap. 6). A wall with tieback anchors, in contrast, would result in a very different stress distribution because of the large compressive forces applied to the soil mass and to the failure surface by the anchor system (Fig. 1b). The situation illustrated in Fig. 1b may well preclude the development of progressive failure and may result in other, equally significant, changes in the way testing and design should be done. Additional discussion of this point is presented later in this paper, but, unfortunately, the necessary research and field observations to quantify the example are not available.

Soft Saturated Clays and Silts

The similarity in behavior of soft saturated clays and silts stems from the low permeability of these soils. Because of this low permeability, and in contrast to granular soils, undrained and partially drained situations are common in stability problems involving these materials. Important time-dependent strength changes result when the excess pore pressures associated with undrained stress changes are dissipated through drainage. Consequently, it is common to examine a slope using two limiting bounds of soil strength. At one limit is the undrained shearing resistance, S_u. At the other limit is the completely drained case where shearing resistance depends on the friction angle - $\tau = \sigma' \tan\phi'$.

The characteristic pore pressure response for specimens of soft saturated (normally consolidated or lightly overconsolidated) cohesive soils tested in the laboratory is an increase in pore pressure. However, in field situations involving excavation the overall decrease in total stress may affect the positive pore pressure. Bishop and Bjerrum (1) illustrated the significance of these effects for normal cut slopes (Fig. 2) but the various components of a slope controlled by a wall may be different. Specifically:

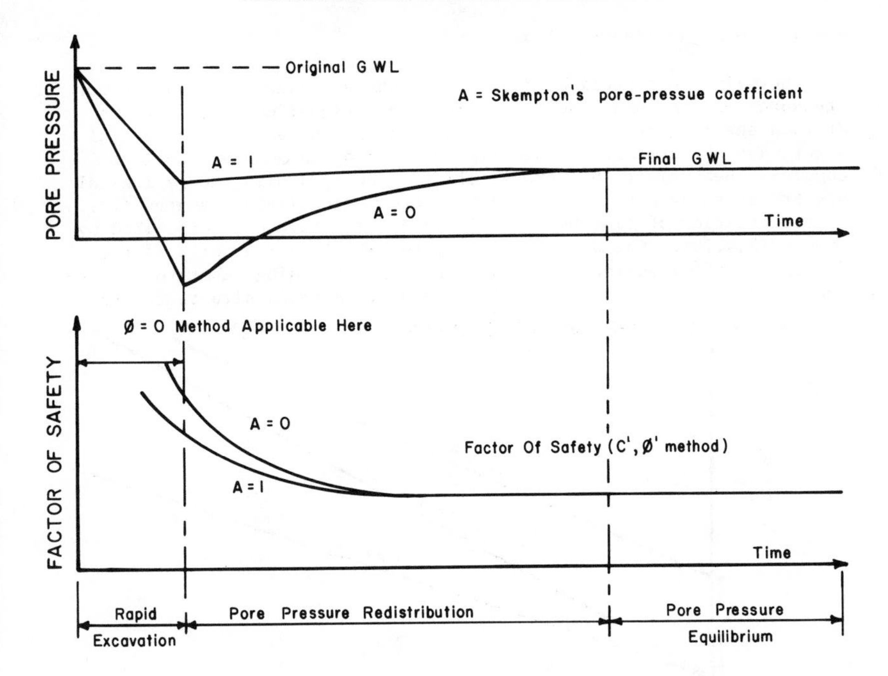

Figure 2. Changes in pore pressure and factor of safety during and after an excavation (after 1).

a. the compressive stresses imposed by the anchor system on the soil mass and failure surface will offset the effects of excavation.

b. the installation of the anchor will generally result in an increase in both the shear stress and normal stress in the soil surrounding the anchor (12)

c. the stresses imposed by the wall on the adjacent soils will be larger than those for slopes without a wall.

d. locking off the tieback load will be a preload on the soil and will limit the potential strain in the soil

In summary, the effects of walls on the stability and shearing resistance of slopes cut in soft silts and clays will be to reduce the potential decrease in shearing resistance associated with drainage. The amount of this reduction or, more appropriately, the level of shear strength available can best be determined through an effective stress analysis (17 - Chap. 7) . The strength parameter, ϕ', for this analysis (c' will almost always equal zero for these materials) can be evaluated using conventional laboratory tests, usually triaxial compression tests.

Heavily Overconsolidated Clay

Heavily overconsolidated clays are characterized by negative pore pressures when sheared under undrained conditions; consequently the drained shearing resistance is lower than the undrained. In several other ways the strength of heavily over-consolidated clays is more complex. Because of the stress relief associated with the geological history of these materials, they are often extensively fissured (18). Characteristics of fissures and fissured soils have been described by others (2,4,19). One problem associated with fissured soils is the difficulty of measuring an appropriate strength using laboratory tests because of size effects (15). As a result, large in situ tests are used

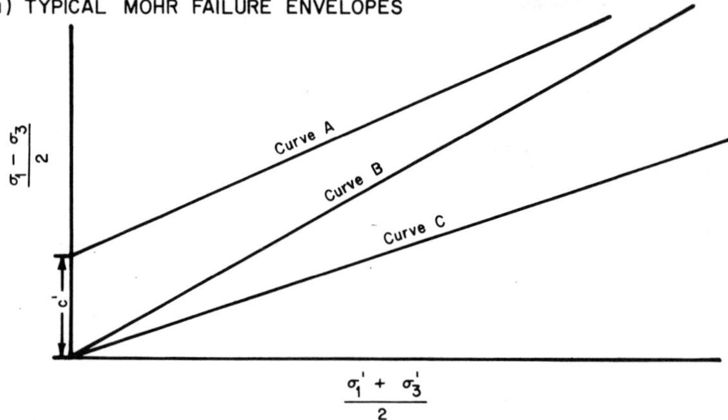

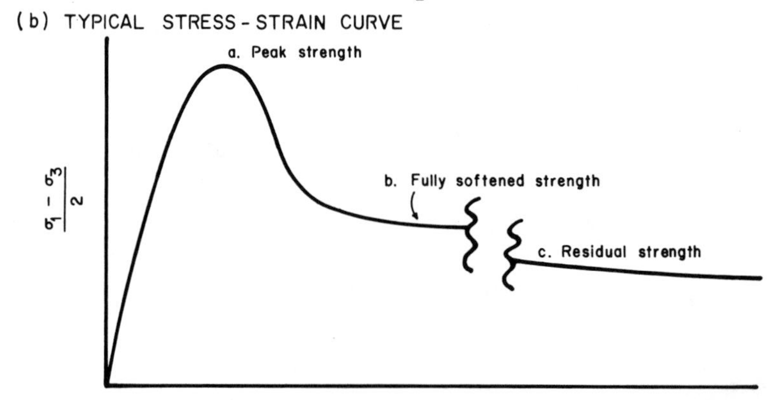

Figure 3. Shear strength levels developed by heavily overconsolidated clays (after 17).

Most heavily overconsolidated clays show strain-softening behavior. A typical stress-strain curve, Fig. 3b, defines several levels of shearing resistance which may be significant in analysis of landslides or walls controlling landslides. When the values of peak strength (point a) from a series of tests on similar specimen are used to define a failure criterion, this relationship will define a cohesion intercept, c', if extrapolated to the axis of $\sigma' = 0$ (Curve A in Fig. 3a). Laboratory tests to define strength should be done using the same level of normal stress expected in the field because research (2,18) has shown that this failure envelope is curved in the low stress region and passes through the origin. The stress changes associated with anchored walls must be evaluated to interpret the peak strength applicable in these situations.

The second significant level of shearing resistance in heavily overconsolidated clay is the fully softened strength (point b). The failure envelope at this limit (Curve B in Fig. 3) is linear and passes through the origin with $c' = 0$ and ϕ' essentially the same as for normally consolidated clay of the same type (17 – Chap. 6). The fully softened strength is applicable to many slope stability problems (6, 20) and may be particularly important in situations where walls limit the amount of deformation in a slope. Under these conditions progressive failure may be limited and the strains required to develop residual strength may not develop.

Residual strength (point c) and the failure criterion defined by the residual friction angle, ϕ'_r[2], develop only after very large deformations (9,16). The value of the residual friction angle may be significantly lower than the peak or fully softened value of ϕ' (more than 10° in some soils) and is dependent upon the soil mineralogy.

As a result, the decision as to whether the residual strength should be used in design or not is an important one with large economic implications. To date there is little in research or performance monitoring data to indicate what effects walls might have in limiting the deterioration in strength to the level of the residual strength.

Cohesionless Soils

For granular soils, such as gravel, sand and nonplastic silts, the failure envelope is approximately linear and passes through the origin defining $c' = 0$. In most applications, sandy and gravelly soils of high permeability are able to drain in response to changes in applied stress and pore pressure. An effective stress analysis can be done using the position of the groundwater to define the pore pressures. Effective stress strength parameters are used.

Laboratory evaluation of strength parameters for granular soils is usually very straightforward because tests can be done under drained conditions. Both triaxial compression tests and direct shear tests are common. Interpretation of test results is sometimes complicated because the failure envelope is curved at high normal stresses or because of higher values of ϕ' measured in direct shear compared with triaxial tests (17 – Chap. 6).

Other Types of Material

The brief review above has been limited to a highlighting of major principles. Many other characteristics of soil and rock strength may be important in a specific situation. Among these characteristics are soil sensitivity, response to cyclic loading, partially saturated soil response and others which have proven to be important in specific cases. The strength of rock masses, weathered rock, residual soils and colluvium is a key to many landslide problems. For all of these subjects there is an extensive scientific literature.

DEFORMATION PROPERTIES

The principal application of deformation properties in engineering control of landsliding is the use of stress-strain relationships in finite element analysis. Although this method is not widely applied at present, impressive contributions have been made and it is reasonable to expect more widespread use in the future. The most challenging and rewarding applications are likely to be in soil-structure interaction problems, such as the use of walls to control landslides (17 - Chap. 8).

Successful finite element analysis depends on realistic constitutive models for the materials involved. Models for soil masses have ranged from linear to very complex nonlinear forms. Hyperbolic models (7) are popular and have been incorporated in many commercial programs.

Another important aspect of finite element analysis is modeling of interfaces between soils and other materials. Failure around anchors, contact between walls and soil and the surfaces within jointed and fractured rock are all important problems where interface models will dominate the finite element analysis. Some work has been done on this subject (11) but there are few descriptions of applications to control of landslides using walls.

In any testing program to evaluate parameters for finite element analysis applied to soils, several principles must be applied (14):

a. high quality undisturbed test specimens should be used at all times to model natural soils. Compacted soils and interfaces should be prepared to match field conditions.

b. the initial stress conditions, specifically anisotropic conditions, from the field should be reproduced in the laboratory.

c. the level of stress used in laboratory testing should be the same as in the field or scaling techniques such as SHANSEP (13) should be used.

d. the strain constraint from the field, (e.g., plane strain) should be consistent in the laboratory equipment.

SIGNIFICANCE OF GROUNDWATER

Groundwater is a major factor in almost every landslide. Both the seepage forces associated with groundwater flow and the reduced shearing resistance available when a rise in groundwater lowers effective normal stresses, contribute to instability. The significance of this factor is equally important in engineering control of landslides using walls.

In most cases the inherent seasonal variation in the level of groundwater introduces the complexity of time variability. This variability is expressed in both a large uncertainty and a probable bias with respect to any physical measurements used to evaluate the position of the phreatic surface. To illustrate this point, it is typical to evaluate the location of groundwater during an engineering assessment of landsliding by installing one or more observation wells and recording the level of water in these wells. Unless wells existed at the site beforehand, data will be available only for the period of observation which is rarely more than one or two years, and usually less than one year. The characteristic being measured, groundwater level, will vary significantly over a long period of time as well as within one season. It may be meaningful to talk about groundwater levels in terms of probability of recurrence in twenty-year or fifty-year design periods, as is common in stream flood flows. In this case the significance, or insignificance, of data from a short period of well observations is clear. Since the critical design condition for a slope will be the highest level of groundwater expected within the project lifetime, it is this groundwater level which should be predicted. The groundwater data collected from an observation well during a short and arbitrary period may have little bearing on the levels causing failure during the project lifetime. There is no reasonable or rational way to simply increase the factor of safety or add an increment of additional groundwater as a safety measure. The costs and risks involved make this approach unacceptable, although it is used.

Predicting Groundwater Levels

Typical data illustrating the relationship of precipitation to landsliding are illustrated in Figure 4. The link between these two events is groundwater level which will fluctuate with variation in precipitation and might be influenced by surface irrigation, dewatering systems and other man-made factors. The natural geological details of a particular site are also important in establishing a relationship between precipitation and landsliding. Such factors as a perched water table, anisotropic medium or a complex geologic structure will control the specifics of groundwater level at a particular site. A rational prediction of landsliding resulting from precipitation must address all of these factors.

Methods are available for making a rational prediction of groundwater level and its fluctuation (10). A first step is to define the separation of total precipitation into components associated with runoff, evapotranspiration and effective recharge (Fig. 5). It is the recharge which produces change in the groundwater levels.

Consequently, much better correlation can be expected between the incidence of stability problems and recharge than with the total amount of precipitation.

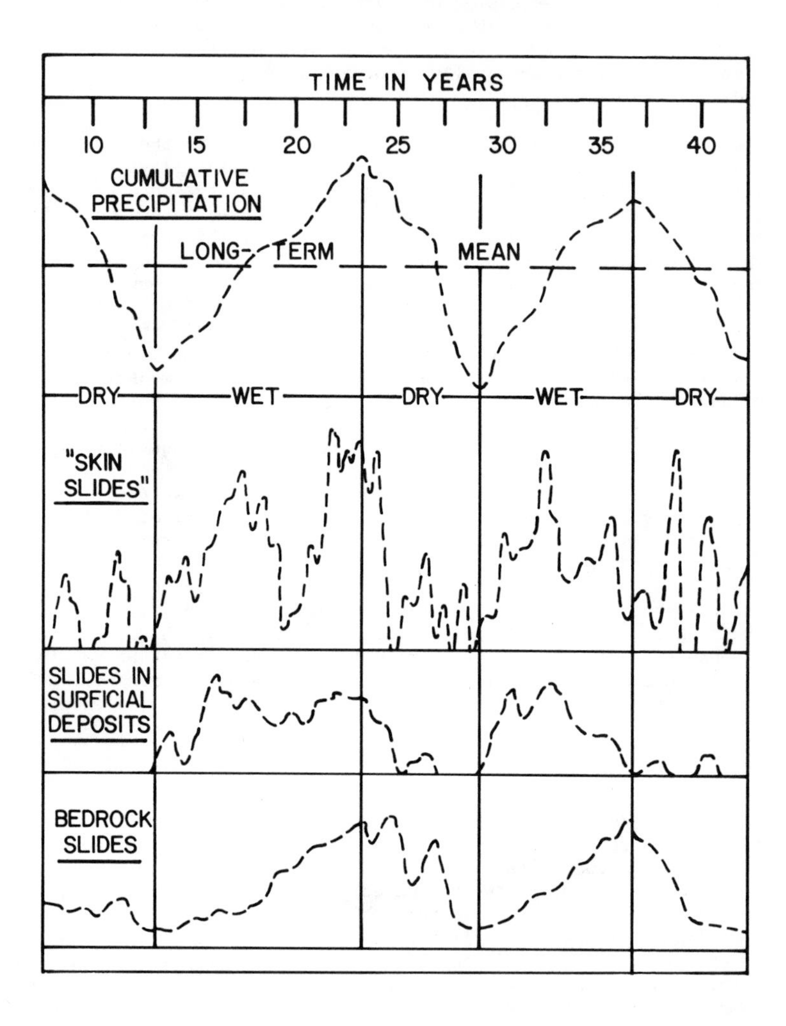

Figure 4. An illustration of the different relationships between landsliding and precipitation which produces changes in groundwater level (after 5).

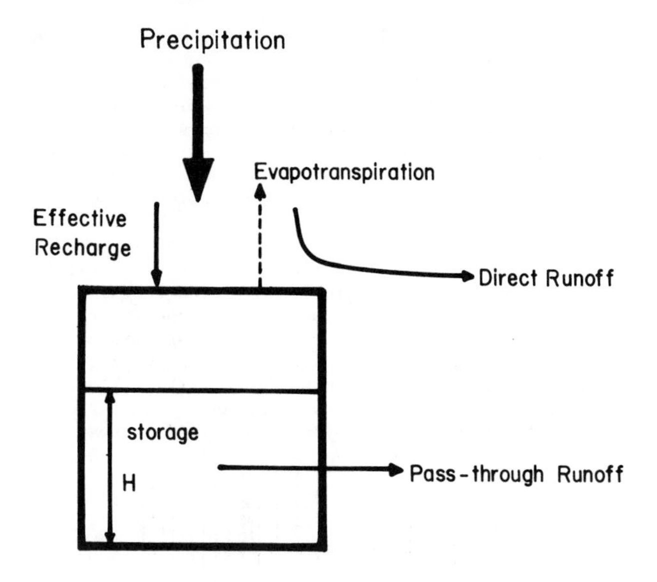

Figure 5. Groundwater flow model illustrating the effective recharge.

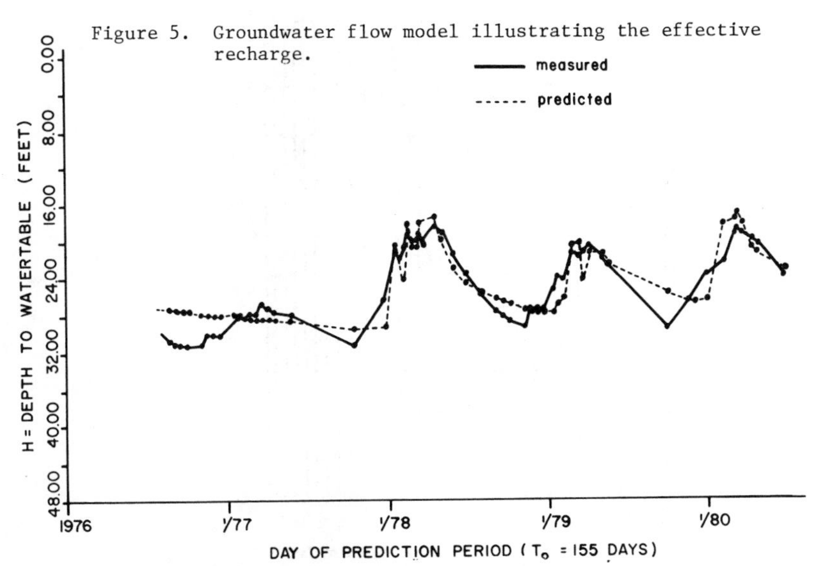

Figure 6. Illustration of the prediction of groundwater fluctuation based on effective recharge and precipitation records.

One newer technique (10) which has been used successfully to predict the relationship between precipitation, groundwater level and stability problems uses a site specific weighting function to be multiplied by each increment of effective recharge. The weighting function is large for recent rainfall and decreases to smaller values for rainfall occurring in the past. When calibrated with site specific well data, predictions using this method are quite accurate (Fig. 6) and can be used as the basis for rational long-term predictions of groundwater levels.

An application of this model is to use existing rainfall records from a particular area, along with calibration from a specific site, to simulate the groundwater levels during the past period of the rainfall record. Having done this simulation, a statistical description of the data can be developed; for example, the recurrence interval for various levels of groundwater can be defined (Fig. 7). If this kind of information is available, then design or analysis of slopes, including control works for slopes, can be done in a rational way. The concept

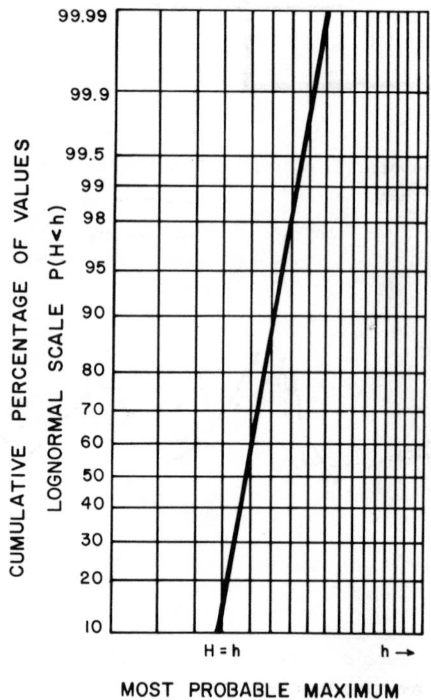

Figure 7. Probabilistic description of groundwater level for use in design.

of designing for a particular risk level or for a particular recurrence interval is preferable to arbitrary factors of safety applied to a single measured groundwater level.

When walls are used to control landsliding, it is usually feasible to construct drainage control at the same time. When groundwater levels can be predicted, as in Figure 7, the design of groundwater control systems, such as horizontal drains, can be based on the drainage requirements anticipated over the lifetime of the landslide control project.

CONCLUSIONS

Evaluating the properties of landslides controlled by walls presents many of the same problems as those associated with uncontrolled landslides. The principles of effective stress methods for defining soil strength and deformation properties must be applied to designs when time or drainage effects are involved. The range of soil properties can be anticipated based on an extensive literature on the subject.

Introduction of a wall, especially a tieback wall, will change the stress distribution within the slide mass and may limit the amount of deformation. Consequently, some important mechanisms of soil failure in uncontrolled slopes, such as progressive failure and development of residual strength, may not be as significant when walls are used.

The expected level of groundwater in a slope can be predicted for the lifetime of the control project. Design or control strategies should be based on these predicted levels.

REFERENCES

1. Bishop, A.W., and Bjerrum, L. The Relevance of the Triaxial Test to the Solution of Stability Problems. Proc., Research Conference on Shear Strength of Cohesive Soils, Boulder, American Society of Civil Engineers, New York, 1960, pp. 437-501.

2. Bishop, A.W., Webb, D.L., and Lewin, P.I. Undisturbed Samples of London Clay From the Ashford Common Shaft. Geotechnique, Vol. 15, No. 1, 1965, pp. 1-31.

3. Bjerrum, L. Progressive Failure in Slopes of Overconsolidated Plastic Clay and Clay Shales. Journal of Soil Mechanics and Foundations Division, American Society of Civil Engineers, New York, Vol. 93, No. SM5, 1967, pp. 3-49.

4. Brooker, E.W., and Ireland, H.O. Earth Pressures at Rest Related to Stress History. Canadian Geotechnical Journal, Vol. 2, No. 1, 1965, pp. 1-15.

5. Campbell, R.H. Soil Slips, Debris Flows, and Rainstorms in the Santa Monica Mountains and Vicinity, Southern California. Geol. Survey Prof. Paper 851, USGS, 1975, pp. 51.

6. Chandler, R.J. Lias Clay: The Long-Term Stability of Cutting Slopes. Geotechnique, Vol. 24, No. 1, 1974, pp. 21-38.

7. Duncan, J.M., and Chang, C.Y. Nonlinear Analysis of Stress and Strain in Soils. Journal of the Soil Mechanics and Foundations Division, ASCE, Vol. 90, No. SM5, 1970, pp. 1629-1653.

8. Hanna, T.H. Foundation Instrumentation. Trans. Tech. Publications, Cleveland, OH, 1973.

9. Kenney, T.C. The Influence of Mineral Composition on the Residual Strength of Natural Soils. Proc., Geotechnical Conference on Shear Strength Properties of Natural Soils and Rocks, Norwegian Geotechnical Institute, Olso, 1967, Vol. 1, pp. 123-129.

10. Klaiber, J.A. Predicting Landslide Hazards Associated with Precipitation-Related Groundwater. Report R-81-131, Department of Civil Engineering, Carnegie-Mellon University, Pittsburgh, 1981.

11. Kulhawy, F.H., and Peterson, M.S. Behavior of Sand-Concrete Interfaces. Proceedings, Sixth Pan American Conference on Soil Mech. and Found. Engng., Lima, Vol. 2, Dec., 1979, pp. 225-236.

12. Kulhawy, F.H., Sangrey, D.A., and Clemence, S.P. Uplift Anchors on Sloping Sea Floors - State of the Art. Report M-R510, Naval Construction Battalion Center, Port Hueneme, California.

13. Ladd, C.C., and Foote, R. New Design Procedure for Stability of Soft Clays. Journal of the Geotechnical Engineering Division, ASCE, Vol. 100, No. 667, 1974, pp. 763-786.

14. Ladd, C.C., et al. Stress-Deformation and Strength Characteristics. Proceedings, IX International Conference Soil Mechanics Foundation Engineering, Tokyo, Vol. 2, 1977, pp. 421-494.

15. Lo, K.Y. The Operational Strength of Fissured Clays. Geotechnique, Vol. 20, No. 1, 1970, pp. 57-74.

16. Morgenstern, N.R., and Tchalenko, J.S. Microscopic Structures in Kaolinite Subjected to Direct Shear. Geotechnique, Vol. 17, No. 4, 1967, pp. 309-328.

17. Schuster, R.L., and Krizek, R.J. Landslides: Analysis and Control Spec. Rpt. 176, Transportation Research Board, National Academy of Sciences, Washington, D.C. 1978.

18. Singh, R., Henkel, D.J., and Sangrey, D.A. Shear and K_o Swelling of Overconsolidated Clay. Proc., 9th International Conference on Soil Mechanics and Foundation Engineering, Moscow, Vol. 1, Part 2, 1954, pp. 143-147.

19. Skempton, A.W. The Pore-Pressure Coefficients A and B. Geotechnique, Vol. 4, No. 4, 1954, pp. 143-147.

20. Skempton, A.W. First-time Slides in Over-Concolidated Clays. Geotechnique, Vol. 20, No. 3, 1970, pp. 320-324.

TIEBACKS USED FOR LANDSLIDE STABILIZATION

by

David E. Weatherby[1], M. ASCE and Peter J. Nicholson[2] M. ASCE

INTRODUCTION

A variety of techniques can be used to stabilize a landslide or prevent a landslide from developing. Among the most common are:

1. Concrete retaining wall.
2. Buttress fill.
3. Reinforced earth wall.
4. Excavation of the sliding soil.
5. Relocation of the structure.
6. Regrading.
7. Cantilevered wall.
8. Soil reinforcement.
9. Drainage.
10. Tiedback wall.

For a major slide, the first seven alternatives usually will require either acquisition of additional property above or below the slide area, closing of the facility, or both. The last three measures can usually be performed with a minimum of disruption to existing services, and they often are less expensive than the other methods.

Tiebacks can effectively stabilize most existing landslides and prevent many landslides from occurring, because they can penetrate the failure surface and apply the force required to stabilize the sliding mass.

DESIGN

A tieback is a structural element which uses a grouted anchor in the ground to secure a tendon which applies a force to a structure. Figure 1 shows the components of a tieback. The anchor length is the portion of the tieback where the force is transmitted to the ground. The portion of the tendon between the anchor and the structure is not bonded to the ground, and it is free to elongate elastically. Force is applied to the tieback by post-tensioning.

---

(1) Vice President, Schnabel Foundation Company, 4720 Montgomery Lane, Bethesda, MD 20814.

(2) President, Nicholson Construction Company, P. O. Box 98, Bridgeville, PA 15017.

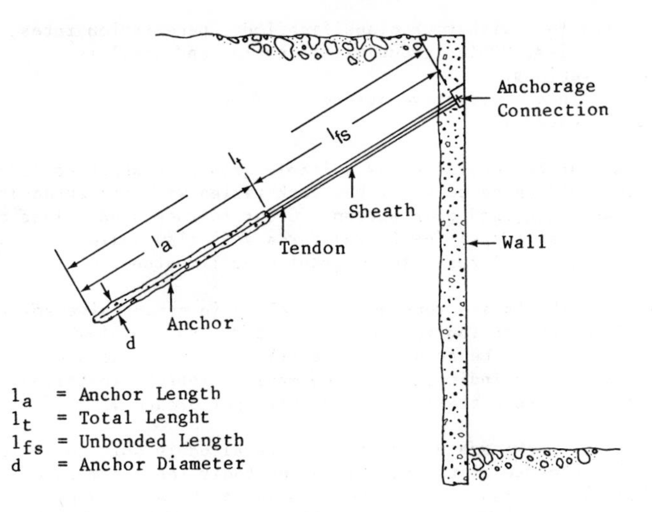

FIG. 1. Components of a Tieback.

An initial evaluation is made to determine whether or not tiebacks can be used at a particular site, and if they will be able to develop the necessary capacity without excessive movement or loss of capacity with time. Tieback capacities are estimated using empirical relationships developed for the particular tieback type.

The soil or rock properties in the vicinity of the tiebacks are required in order to determine if they can be used. The tieback contractor also requires soils information in order to select the tieback type, and estimate their capacities.

If the tieback is to be made in soil, the following properties or test results should be obtained and included in the contract documents:

1) Boring logs with standard penetration resistances, visual classifications, groundwater levels, and drillers observations.
2) Unified Soil Classification System descriptions of the soil.
3) Plastic and liquid limits.
4) Unconfined compressive strength on undisturbed and remolded clay samples.
5) Grain size distribution curves for fine grained sands and silts.
6) Resistivity, soil and groundwater pH, soluble sulfate content, and sulfide content are necessary for corrosion protection selection.

If the tieback is to be located in rock, the following properties or test results should be included in the contract documents:

1) Boring logs with rock classifications, penetration rates, recoveries, RQD's, groundwater levels, and drillers observations.
2) Unconfined compressive strength.
3) Groundwater pH.

All rocks can be considered as suitable ground in which to found anchors. The holding capacity of the rock varies with its structure, compressive strength, and composition. Grout to rock bond values can vary from as low as 10 psi (69.0 kPa) for a soft clay shale to as high as 300 psi (2070.0 kPa) for a sound granite or limestone.

Permanent tiebacks are routinely installed in noncohesive soils with a standard penetration resistance greater than ten blows per foot. They should not be anchored in a fill. Testing and monitoring of many installations indicates that permanent tiebacks installed in sandy soils will have satisfactory long-term performance.

Permanent tiebacks are not routinely installed in soft to medium cohesive soils because of the concern about their long-term load holding ability. In many cases, these soils can be avoided by drilling the tiebacks at a steeper angle and to a depth where better soil or rock may be found. Tiebacks installed in soils with a high organic content, normally consolidated clays, and cohesive soils with an unconfined compressive strength less than 1.0 ton/ft^2 (96 kPa) and remolded strengths less than 0.5 ton/ft^2 (48 kPa) may be creep susceptible. Tiebacks installed in soils that exceed these strengths, and have a consistency index (I_c) [1] greater than 0.8 have not experienced significant loss of load or movement with time. The consistency index is a given by the relationship:

$$I_c = \frac{W_L - W}{W_L - W_P}$$

W_L = Liquid limit
W = Natural water content
W_P = Plastic limit

If permanent tiebacks are to be anchored in a cohesive soil, or sandy soil with a standard penetration resistance less than ten blows per foot, then a testing program should be used to evaluate the long-term load carrying capacity of the tieback.

When using permanent tiebacks to support a retaining wall, the following items will serve as a guide:

1. Design Load Between 50 and 130 tons (445 and 1156 kN) - Tiebacks tendons of this capacity can be handled without the need of heavy equipment and the drill hole size need not be larger than 4 inches (102 mm). In addition, the stressing and testing equipment can be readily handled without using power lifting equipment.

2. Length - Due to stability requirements, there are few retaining wall tiebacks that are shorter than 40 feet (12.2 m). A minimum unbonded length of 15 feet (4.6 m) should be adopted to avoid unacceptable high prestress losses due to long-term relaxation and creep in the prestressing steel and soil, and anchorage seating losses. The unbonded length of the tieback should be selected so that the anchor is located beyond the potential critical failure surface. The unbonded length rarely exceeds 150 feet (45.75 m). The total tieback length should be established so the most probable failure surface passing through the ends of the tiebacks or behind them would have a factor of safety equal to or greater than the factor of safety on the critical failure surface. Figure 2 schematically illustrates these two cases.

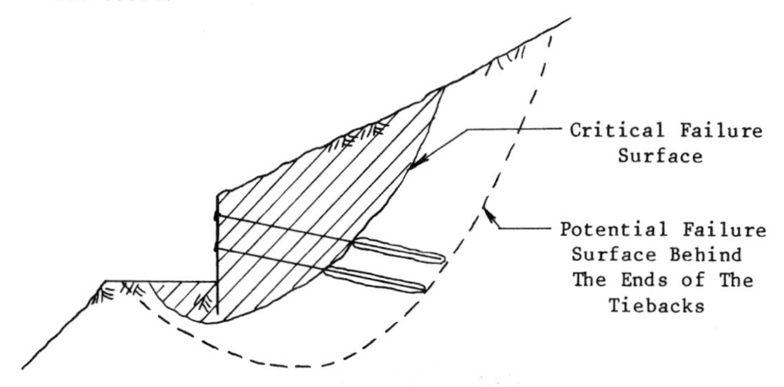

FIG. 2. Stability Analysis for Determining the Unbonded and the Total Tieback Length.

3. Tieback Locations - If the failure surface is shallow and the slide prevention wall penetrates the failure surface, then one row of tiebacks can be used to support the wall. Figure 3 schematically shows such a wall. In this case, the wall may be designed to provide shear resistance across the failure surface. When the failure surface is deep, then the wall can be designed to stabilize the landslide without requiring the wall to penetrate the failure surface. In this case, at least two rows of tiebacks are required to provide a stable wall. Figure 4 shows a multirowed tiedback wall which did not penetrate the failure surface. If the sliding mass has continuity (weathered rock or rock), then a wall may not be necessary, and individual elements, buttresses, or blocks can be tiedback. Figure 5 shows anchored concrete blocks protecting a tunnel portal and preventing a rock slide.

4. Angle of Inclination - It is desirable that a minimum of 15 feet (4.6 m) of overburden be above the anchor. Most soil tiebacks are installed at an angle of between 10° and 30°. Special grouting techniques may be required if the angle of inclination is less than 10°. When a suitable

anchoring strata lies at some depth, generally more than 30 feet (9.2 m), an angle of 45° may be chosen as a compromise between total tieback length and decrease in the resultant horizontal force for a given tieback capacity. It must be kept in mind, though, that by increasing the angle of inclination, the vertical component of the tieback load also increases, thus increasing the vertical load on the wall members and the underlying foundation material.

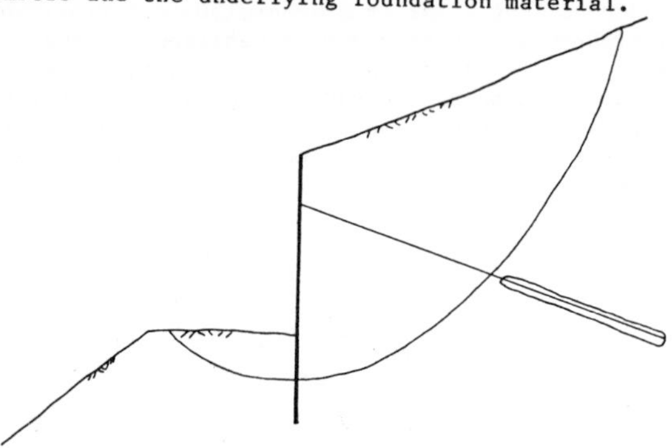

FIG. 3. One Tiered Tiedback Wall.

FIG. 4. Multitiered Tiedback Wall (Courtesy Karl Bauer)

FIG. 5. Anchored Blocks (Courtesy VSL Corp.)

5. Drilled Hole Diameter Between 3 and 6 Inches (76 and 152 mm) - The vast majority of soil tieback work is performed using a cased hole. The weight of the casing and associated handling and drilling problems related to larger casings presently makes 6 inches (152. mm) the largest size in common use. Most common casing sizes are 3 inch (76 mm) O.D. (used with percussion methods) and 5 inch (127 mm) O.D. (used with rotary drills).

6. Tendons - The two types of tendons normally used are:

   a. Seven-wire strand of 270 ksi (1.86 MPa) ultimate tensile strength, either 0.5 or 0.6 inches (12.7 or 15.2 mm) in diameter conforming to ASTM 416 specifications.
   b. Deformed bar of 150 ksi (1.03 MPa) ultimate tensile strength, usually 1.0, 1.25, or 1.375 inches (25.4, 32.0, or 36.0 mm) in diameter and conforming to ASTM 722, Type II specifications.

7. Easements - The owner must be able to obtain tieback easements from adjacent property owners.

The most economical tieback installation will be obtained if the design enables the contractor to select the tieback type, the construction method, and the tieback capacity. The designer should specify the minimum unbonded length, the minimum total tieback length, and the unit tieback capacity required at each tieback level. In lieu of specifying a unit capacity a loading diagram could be specified, and the contractor would design and furnish the complete wall. In addition, the type or desired level of corrosion protection must be specified as well as the method of verifying the long-term load carry capacity. Finally, each production tieback should be tested to verify that the anchor will carry the design load.

CORROSION PROTECTION

Permanent tiebacks have been installed routinely since the mid-1960's in Europe, and since the early 1970's in the United States. They are performing well in a variety of environments. Most tiebacks use cement grout for protection over their anchor length. Portier [2], and Herbst [3] reported that there is no evidence of a corrosion failure where the tieback tendon was encased in grout. Corrosion failures have occurred along the unbonded length of unprotected tendons, with most of them located within 6.65 feet (2 m) of the anchor head. A significant number of the tieback corrosion failures in Europe occurred in tendons fabricated using quenched and tempered prestressing steels. These steels do not meet ASTM specifications,and they have not been widely used in the United States.

Most permanent tiebacks can be protected by portland cement grout along the anchor length, and a grease filled tube or heat shrinkage sleeve over the unbonded length, Figure 6. Grout protected tiebacks should be electrically isolated from the structures they support, and the tendon should have a minimum of 0.375 inches (9.5 mm) of grout protection. Figure 6 shows the anchorage insulation used to isolate the tendon. Electrical isolation interrupts the long-line differential aeration corrosion cell shown in Figure 7. This cell is potentially dangerous because it does not require oxygen in the soil, and because of the relative size of the cathode and anode. If this cell develops, the tendon at the top of the anchor zone would become the anode, and the entire wall would become the cathode. Electrical isolation also would interrupt any stray current corrosion system.

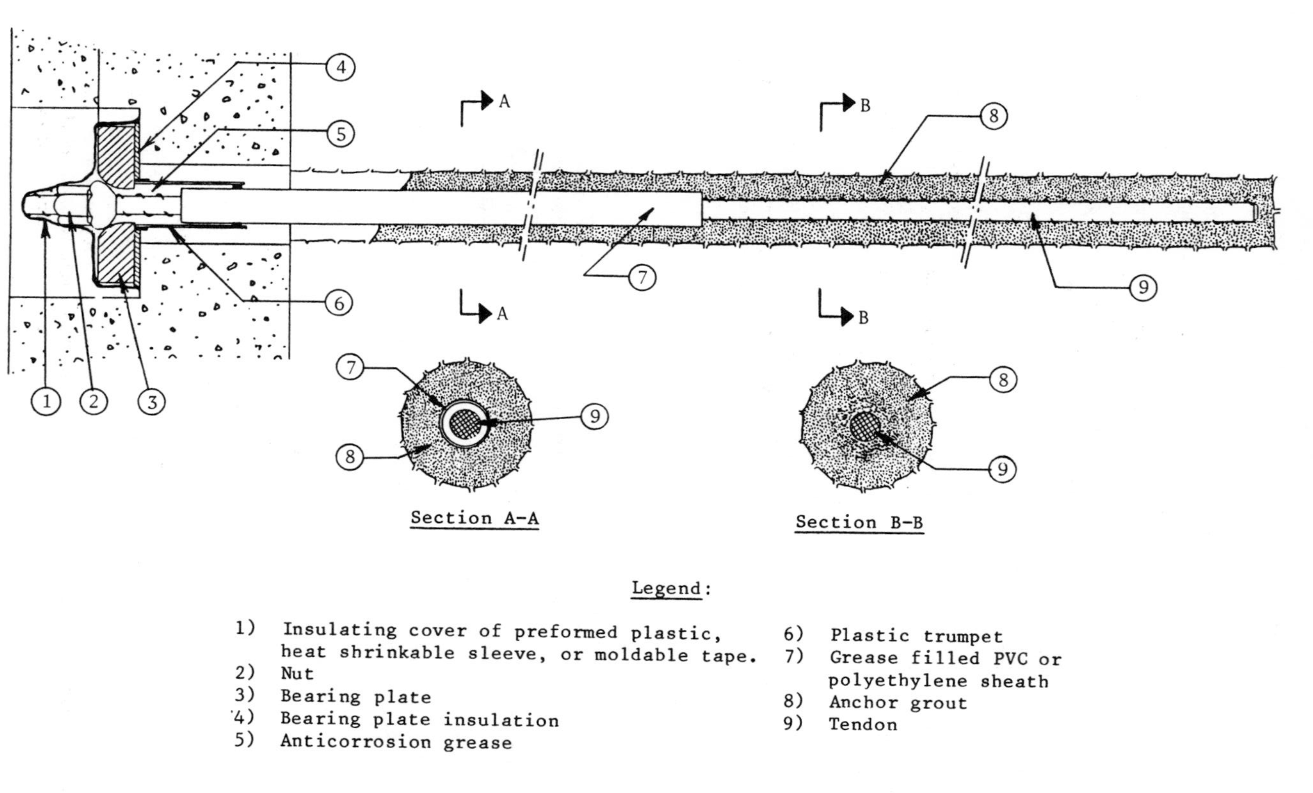

FIG. 6. Grout Protected Tieback.

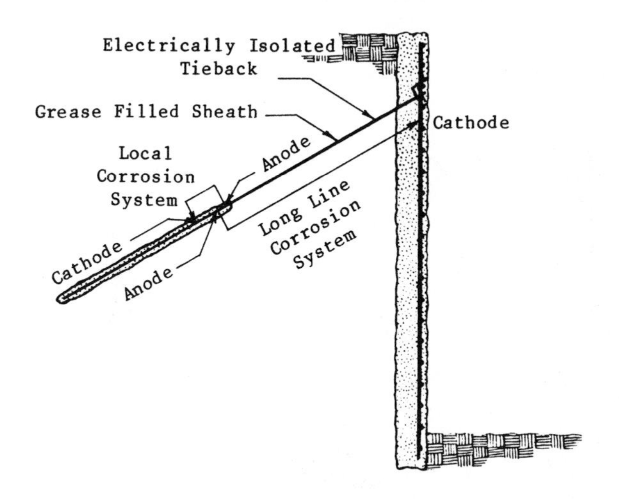

FIG. 7. Corrosion Mechanisms Affecting a Tieback.

If the soil surrounding the anchor length has a pH less than 5.0, or a resistivity less than 2,000 ohm-cm, or sulfides present, then a local corrosion system could develop on the tendon, Figure 7. If these conditions exist, then the tendon should be completely encapsulated in a plastic or steel tube. Figure 8 shows an encapsulated tieback. The encapsulation will interrupt any long-line and stray current corrosion system, and prevent the local corrosion system from developing.

Figures 6 and 8 show two ways to provide corrosion protection for the anchorage, and the tendon below the bearing plate. Care must be taken to insure that this area is well protected since most known corrosion failures have occurred near the anchor head. The corrosion protection under the anchorage should be designed to accommodate small movement of the wall.

The American Water Works Association [4] describes how the pH, resistivity, and sulfide content can be measured in the field. The soluble sulfate content of the soil should be determined in the laboratory. If the soluble sulfate content exceeds 2,000 mg/kg, then ASTM Type V cement should be used. When the pH of the soil is less than 5 or when buried concrete structures in the vicinity are suffering from acid attack, portland cements should not be used for the anchor grout.

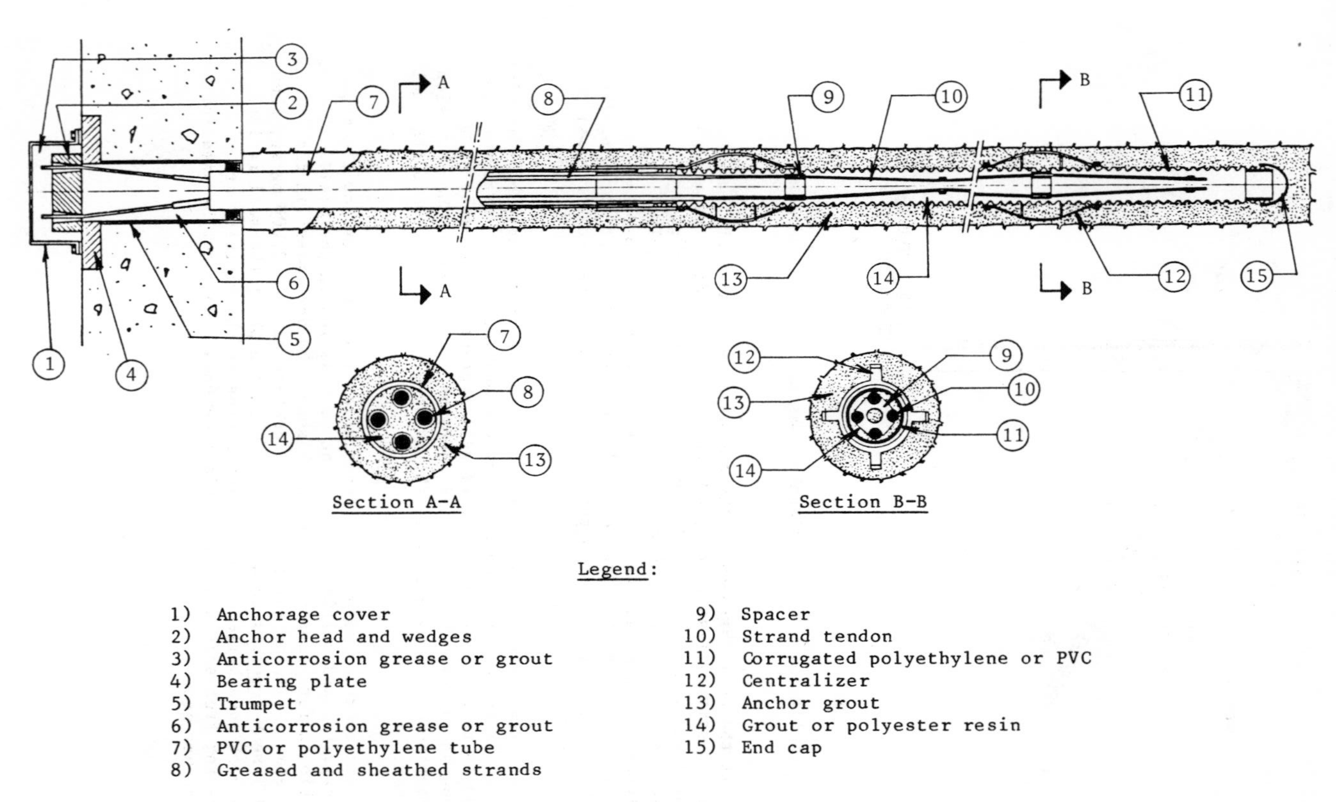

FIG. 8. Encapsulated Tieback

SPECIFICATION

The specification should establish a quality level without eliminating suitable proprietary tieback systems or methods. The designer may require the prequalification of the tieback contractor. The prequalification can be based on experience, or a list of acceptable contractors could be included in the specifications.

An alternative type of prequalification warrants evaluation. This method would require the submission and approval of the tieback system, and the corrosion protection prior to bid. This procedure was used at the Edmonton Convention Center. The submission must be detailed enough to enable the designer to determine if his design is satisfied. This form of prequalification would also allow the contractor to know if his proprietary techniques would be acceptable, and the owner would be able to take advantage of any cost savings. Preparation and review of the submittal would not require a great deal of time, and this contracting procedure would encourage alternate tieback types, and continued tieback development.

CONSTRUCTION AND GENERAL CONSIDERATIONS

There are a variety of construction methods that can economically be used to install tiebacks at the same site. Tiebacks can usually be installed with a minimum of disruptions to existing services. Access, soil or rock type, tieback length, tieback capacity, corrosion protection requirements, and properitary methods will all affect the selection of the drilling and the grouting method.

The vast majority of permanent tiebacks use neat-cement grout made with portland Type I, II, or III cement and water. Additives should be avoided unless there is an overriding engineering reason for using them. Expansion agents that release a gas are to be particularly avoided as they tend to weaken the grout and make it more permeable. Any grouts containing chlorides should not be used as they may cause corrosion of the steel tendon. Likewise, grouts with thixotropic properties or anti-bleed agents have been found to reduce bond.

If the rock is broken and fractured as often happens in a side hill cut where past stress relief has affected the rock, care must be taken to prevent loss of grout from around the tendon. There are two ways to prevent this from occurring.

1. Where a greased and sheathed tendon is used, the grouting of the tieback can be carried out in a single stage. Single stage grout is tremied into the hole from the bottom of the hole until good grout is seen to flow out of the top of the hole. Observation of the grout level for 1/2 to 1 hour to see that it does not fall more than a few feet will then ensure that the anchor grout is remaining in place.

2. Preliminary grouting and then redrilling of the anchor zone. This should insure the sealing of the rock. As a check it is useful to see if the hole will hold water. A water loss of

more than 2 or 3 gal/min (1.26 or 1.89 X 10^{-4} m^3/s) in a 4 inch (102 mm) diameter hole 20 feet (6.1 m) long indicates addition grouting may be necessary.

When a fill is to be placed around the unbonded length of a tieback, care must be taken to protect the tendon during the backfilling operation. The backfill should be hand placed and tamped around the tendon. In addition, a steel pipe may be used to protect against damage by backfilling equipment. If the backfill is expected to settle significantly, then strand tendons should be used because they can tolerate bending better than a bar.

TIEBACK TESTING and MONITORING

Every tieback should be tested to verify that it will carry the design load, without excessive movement, for the service life of the structure. Tiebacks are one of the few structural systems where every member can normally be tested before placing them into service. Three types of tests are used; performance, proof, and creep tests.

A hydraulic jack and pump are used to apply the load. The entire tieback tendon should be simultaneously loaded during testing. The movement of the tieback is measured with a dial gauge or a vernier scale supported on a reference which is independent of the tiedback structure. Movement cannot be accurately monitored by measuring jack ram travel.

The first few tiebacks and a selected percentage of the remaining tiebacks should be performance tested. The performance test is used to establish the load deformation behavior for the tiebacks at a particular site. It is also used to separate and identify the causes of tieback movement, and to check that the unbonded length has been provided. The movement patterns developed during the performance test are used as a control to check the results of subsequent proof tests.

Performance testing is done by measuring the load applied to the tieback and its movement during incremental loading and unloading. Table 1 gives the loading schedule for a performance test and contains the results of a test made on a temporary hollow-stem-augered tieback installed in a stiff clayey silt.

Two types of load-movement curves can be plotted for each performance test. Figure 9 (a) shows the total movement curve for the test results contained in Table 1. In order to simplify the presentation of the data and to highlight the behavior of the tieback, only the movement at the maximum load in each increment is plotted. The readings to be plotted are identified with a single asterisk (*) in the remarks column in Table 1.

When a tieback is loaded, the anchor moves through the soil as it develops capacity. When the load is reduced to zero, a portion of the movement is elastic and recovered, but some of the movement is nonrecoverable. This nonrecoverable movement (residual anchor movement), is also measured during a performance test. Figure 9 (b) shows the residual anchor movement curve for the data in Table 1. The residual movements are plotted as a function of the highest previous load and they are identified with double asterisks (**) in the remarks column in Table 1.

TABLE 1. Performance Test Schedule (1.0 ton = 8.9 kN, 1.0 inch = 25.4 mm, 1.0 psi = 6.9 kPa)

Load increment	Basis of load DL=Design Load	Load (tons)	Observation period (min)	Jack pressure (psi)	Movement (inches)	Remarks
0	0	0		0	0	
T0	Alinement Load	5		245	0	*
P1	0.25 DL	17.5		860	0.449	*
T0	Alinement Load	5		245	0.131	**
P1	0.25 DL	17.5		860	0.426	
P2	0.50 DL	31		1525	1.102	*
T0	Alinement Load	5		245	0.203	**
P1	0.25 DL	17.5		860	0.435	
P2	0.50 DL	31		1525	1.097	
P3	0.75 DL	52		2555	1.761	*
T0	Alinement Load	5		245	0.298	**
P1	0.25 DL	17.5		860	0.446	
P2	0.50 DL	31		1525	1.101	
P3	0.75 DL	52		2555	1.778	
P4	1.00 DL	70		3440	2.622	*
T0	Alinement Load	5		245	0.391	**
P1	0.25 DL	17.5		860	0.458	
P2	0.50 DL	31		1525	1.123	
P3	0.75 DL	52		2555	1.787	
P4	1.00 DL	70		3440	2.631	
P5	1.20 DL	84		4130	3.679	*
T0	Alinement Load	5		245	0.762	**
P1	0.25 DL	17.5		860	0.862	
P2	0.50 DL	31		1525	1.523	
P3	0.75 DL	52		2555	2.007	
P4	1.00 DL	70		3440	2.638	
P5	1.20 DL	84		4130	3.689	
P6	1.33 DL	92	1	4525	4.367	*
P6	1.33 DL		2		4.484	
P6	1.33 DL		3		4.529	
P6	1.33 DL		4		4.554	
P6	1.33 DL		5		4.573	
P6	1.33 DL		7		4.593	
P6	1.33 DL		10		4.616	
P6	1.33 DL		15		4.635	
P6	1.33 DL		20		4.646	
P6	1.33 DL		25		4.655	
P6	1.33 DL		30		4.662	
P6	1.33 DL		45		4.680	
P6	1.33 DL		60		4.691	*
P5	1.20 DL	84		4130	4.632	*
P4	1.00 DL	70		3440	4.448	*
Lock-off						

The total movement of a tieback is made up of elastic elongation of the tendon, residual movement of the anchor, elastic movement in the anchor, and creep movements of the anchor and the tendon. These components of elongation can be determined in a performance test, and they are identified in Figure 9.

In noncohesive soils and rocks, the maximum load applied during the performance test is held constant for 10 minutes, and the movements are measured and recorded at the times indicated in Table 1. These tiebacks are not usually creep susceptible, and the movement between 1 minute and 10 minutes has typically been found to be less than 0.04 inches (1 mm). If this is the case, then the test can be discontinued. If the movements exceed 0.04 inches (1 mm) or the anchor is made in a cohesive soil, then the maximum load should be held for 60 minutes. The elongation should be recorded at 0, 1, 2,

3, 4, 5, 7, 10, 15, 20, 25, 30, 45, and 60 minutes, and a representative creep curve then can be plotted. Tieback creep tests and their interruption is discussed at the end of this section.

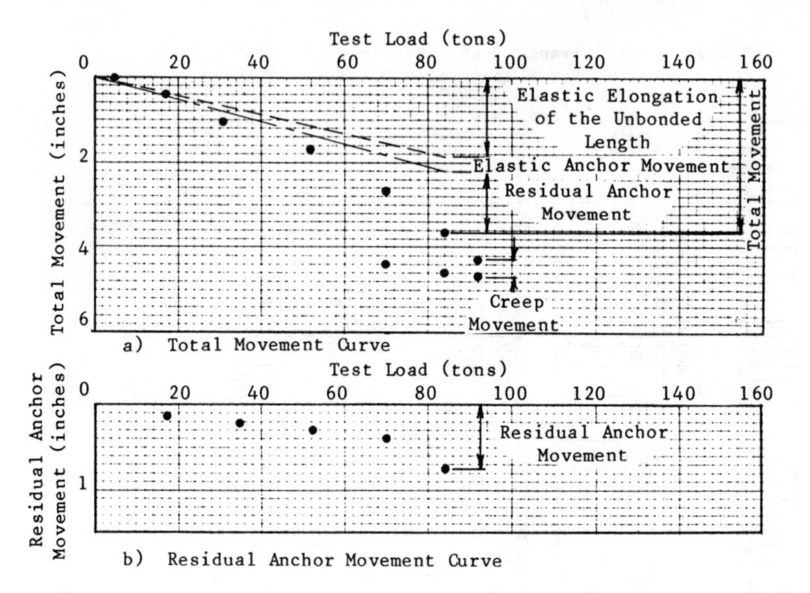

FIG. 9. Performance Test (1.0 ton = 8.9 kN, 1.0 inch = 25.4 mm)

Each production tieback which is not performance tested should be proof tested. A proof test is a simple test which is used to determine the behavior of the tieback. Proof testing is done by measuring the load applied to the tieback and its movement during incremental loading. Table 2 gives the loading schedule for a proof test, and contains the results of a test performed on a hollow-stem-augered tieback. The increments of load are the same as those used in the performance test, except the maximum load may not be as high as the maximum load used in a performance or a creep test. Figure 10 shows a plot of a proof test performed on a hollow-stem-augered tieback. The maximum load applied during a proof test is held constant for 5 minutes and the tieback movement is recorded. If the movement during the 5 minute observation period is less than 0.03 inches (0.76 mm), then the tieback should perform satisfactorily. If the movement exceeds 0.03 inches (0.76 mm), then the load should be maintained until the creep rate can be determined and compared to the creep behavior observed during the performance or creep tests.

Creep tests are performed to evaluate the long-term load carrying capacity of tiebacks installed in cohesive soils. These tests are often conducted during a separate testing program. During a creep test, each increment of load is held constant and the movements are recorded and plotted. Figure 11 shows the plot of creep movement verses time on a semi-logarithmic graph for a postgrouted tieback installed in a stiff clay with a trace of fine to medium sand. Each curve in Figure 11 represents the creep movement for each load increment. The load should be held at each increment until the creep rate is clearly established. Usually the load holds shown in Figure 11 are sufficient to determine the creep rates for permanent tiebacks installed in clay. (The length of the load holds are reduced for temporary tiebacks.) If the creep rate cannot be determined during the observation periods used for Figure 11, then the observation periods should be extended.

TABLE 2. Proof Test Schedule (1.0 ton = 8.9 kN, 1.0 inch = 25.4 mm, 1.0 psi = 6.9 kPa).

Load increment	Basis of load DL=Design Load	Load (tons)	Observation period (min)	Jack pressure (psi)	Movement (inches)	Remarks
0	0	0		0		
T0	Alinement Load	3.5		170	0	
P1	0.25 DL	19		950	0.419	
P2	0.50 DL	38		1850	1.059	
P3	0.75 DL	57		2800	1.781	
P4	1.00 DL	75.5		3700	2.355	
P5	1.20 DL	90.5	1	4450	2.818	
P5	1.20 DL		2		2.835	
P5	1.20 DL		3		2.838	
P5	1.20 DL		4		2.841	
P5	1.20 DL		5		2.845	
P5	1.20 DL		7		2.847	
P5	1.20 DL		10		2.851	
P5	1.20 DL		15		2.856	
P5	1.20 DL		20			
P5	1.20 DL		25			
P5	1.20 DL		30			
P5	1.20 DL		45			
P5	1.20 DL		60			
P4	1.00 DL	75.5		3700	2.680	
Lock-off		57		2800	2.220	

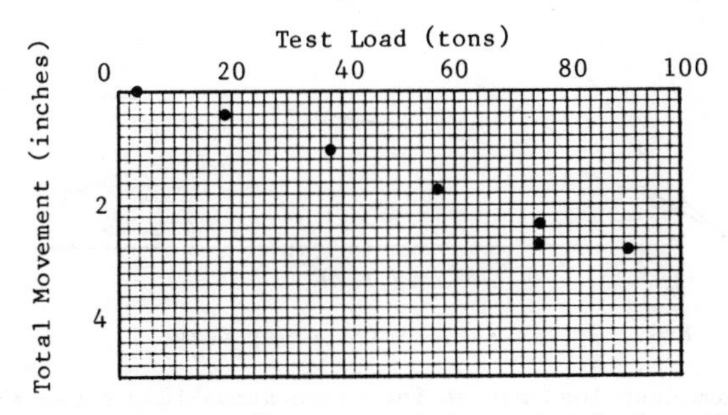

FIG. 10. Proof Test (1.0 ton = 8.9 kN, 1.0 inch = 25.4 mm)

Figure 12 shows the three characteristic types of creep curves observed during tieback testing. Curves (a) and (b) indicated and acceptable behavior as long as the creep movement estimated by projecting the creep rate over the life of the structure is not excessive. For instance, a creep rate of 0.08 inches (2.0 mm) per decade would produce a creep movement of approximately 0.5 inches (12.7 mm) during 50 years. Curve (c) indicates that the tieback would continue to creep until it failed. In the region between curve (b) and (c), it is possible to have a creep curve which slopes gradually upward at the maximum load. This tieback could be accepted, if the creep curve for the design load was similar to curves (a) and (b).

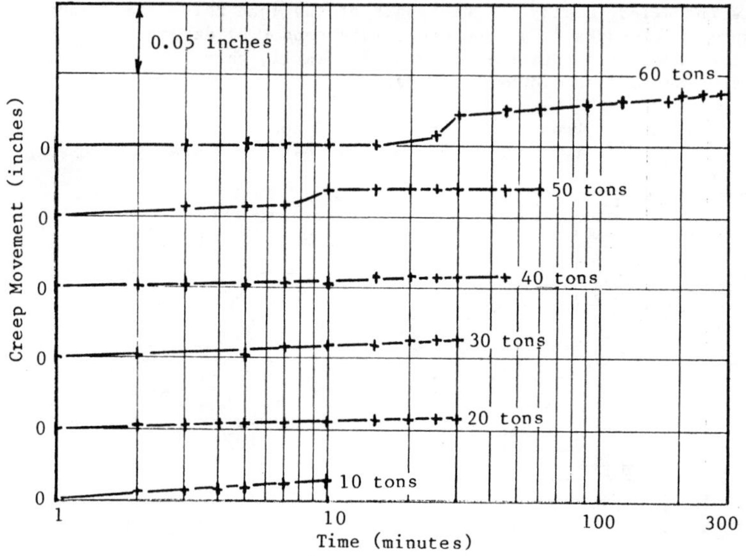

FIG. 11. Creep Test (1.0 ton = 8.9 kN, 1.0 inch = 25.4 mm)

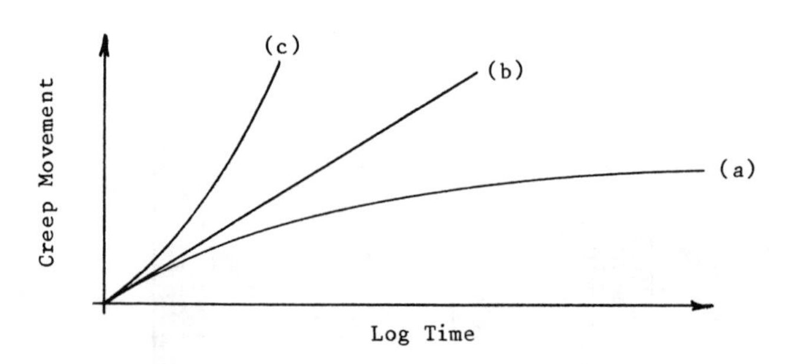

FIG. 12. Characteristic Creep Curves.

The maximum test load may be increased above 1.33 times the design load, depending upon the nature of the soil or the consequences of failure of the structure. In cohesive soils a higher overload reflects the limited experience with permanent tiebacks in clays, and it will cause higher creep movements during testing and reduce the movements after lock-off. The writers are not aware of any long-term performance problems when the tiebacks have been proof tested to 1.20

times the design load, and creep and performance tested to 1.33 times the design load. When it is not possible to establish an independent reference point to measure the movement of each tieback, then a maximum test load between 150 and 200 percent of the design load can be used for the creep and performance tests. Then the remaining tiebacks only need to be stressed and lock-off. When the overload exceeds 150 percent of the design load, then the structure may require reinforcing to resist the test load.

Tieback tests are used to identify the load deformation behavior of each tieback, and to provide data that will enable an engineering evaluation of their adequacy to be made. The total movement curve is used to quickly identify any unusual behavior during a proof test. However, the primary purpose of the test is to verify that the tieback will carry the load without excessive movement. The tieback behavior during the load hold or the creep test provides the best indication of the load carrying ability of the tieback.

Arbitary movement criteria should not be used to establish whether a tieback has passed or failed a test. One type of tieback may require considerable more strain to develop the capacity of the anchor, but once it has been loaded it will perform satisfactorily. Several tieback standards [1], [5], and [6], and many tieback specifications have specified a minimum and a maximum elastic movement for a tieback. The elastic movement of a tieback is equal to the total movement minus the residual anchor movement, and it can only be determined if the load is cycled. Typically the elastic movement has been required to exceed 0.8 times the calculated elastic elongation of the unbonded length, and be less than the calculated elastic elongation of the unbonded length plus half of the anchor length.

The checking of the minimum elastic movement is recommended because it verifies that the unbonded length actually has been provided and that the load has reached the anchor zone. Requiring the maximum elastic movement to be less than the calculated elastic elongation assumes that the skin friction along a straight-shafted tieback is uniform and that the end of the anchor does not move. Measurements and tieback tests have shown that these assumptions may not be true [7] and [8]. Many tiebacks do not transfer load to the soil uniformly, even in uniform soil deposits. If low strength soils are located at the front of the anchor and high strength soils surround the lower portion of the anchor, then the actual elastic movement will exceed the maximum allowed by many of the specifications. Tiebacks that develop a significant proportion of their capacity by bearing also may have elastic movements that exceed those allowed by the standards. When the measured elastic movement exceeds the calculated elastic elongation of the entire tendon, it must be recognized that the entire anchor length has moved through the soil. These tiebacks should not be rejected unless their creep behavior indicates that they would not perform satisfactorily. They should, however, be rigorously tested and monitored before accepting them.

The long-term performance of a tieback can be evaluated by monitoring changes in the tieback load and deformation of the tiedback structure. Both load and deformations must be monitored. The tieback load can be monitored by lift-off tests or load cells. Visual checks, optical surveys, extensometers, or slope indicators can be used to observe deformations.

Tiebacks used for landslide stabilization should be monitored for several years if the failure surface is poorly defined, or if small changes in the soil or rock strengths cause large changes in the estimate tieback force. All permanent tiebacks made in cohesive soils should be monitored.

CONCLUSION

Permanent tiebacks are an economical and effective tools to stabilize or prevent many landslides. They can be installed in rock and noncohesive soils with confidence. Permanent tiebacks can be made in cohesive soils. A rigorous testing program, and a experienced tieback contractor should used if the tiebacks must be anchored in a cohesive soils. The tieback tendon can easily be protected from corrosion, and each tieback should be tested to verify that it will carry the design load for the service life of the structure.

APPENDIX I - REFERENCES

1. Deutsche Industrie Norm, "Verpressanker fur dauernde Verankerungen (Daueranker) im Lockergestein," (Soil and rock anchors; permanent soil anchors; analysis structural design and testing) DIN 4125, Part 2, pp. 1-9, February 1976.
2. Portier, J. L., "Protection of Tie-Backs Against Corrosion," Prestressed Concrete Foundations and Ground Anchors Federation Internationale de la Preconstrainte, Wexham Springs, Slough SL3 6PL England, pp. 39-53, 1974.
3. Herbst, T. F., "Safety and Reliability in Manufacture of Rock Anchors," Presented at International Symposium on Rock Mechanics Related to Dam Foundations, Rio de Janeiro, 1978.
4. American Water Works Association, "Notes on Procedures for Soil Survey Tests and Observations and Their Interpretation to Determine Whether Polyethylene Encasement Should Be Used," Appendix A, AWWA C105-75, 1972.
5. Swiss Society of Engineers and Architects, Ground Anchors, SN 531 191, p. 46, 1977.
6. Post Tensioning Institute, Recommendations for Prestressed Rock and Soil Anchors, PTI, 301 W. Osborn, Suite 3500, Phoenix, AZ 85013, p. 57, 1980.
7. Shields, D. R., Schnabel, Jr., H., and Weatherby, D. E., "Load Transfer in Pressure Injected Anchors," Journal of the Geotechnical Engineering Division, ASCE, GT9, pp 1183-1196, Sept. 1978.
8. Ostermayer, H., and Scheele, F., Research on Ground Anchors in Non-cohesive Soils, Spec. Sess. No. 4, 'Ground Anchors', IX Int. Con. Soil Mech. Found. Engng., Tokyo, Japan, July 1977.

SLIDE CONTROL BY DRILLED PIER WALLS

BY MERLE F. NETHERO[1] M. ASCE

Introduction

Drilled cantilever pier walls have been used in the Ohio River Valley area for the past 15 years in the correction and prevention of overburden cut and fill slope stability problems. This paper summarizes regional experience in the application, design and construction of this form of restraining structure.

Application

Drilled pier walls and the design principles involved have a wide variety of application in providing lateral support. Their primary advantage results from the elimination of the excavation and subsequent backfilling needed when using conventional forms of retaining structure.

Historically, driven timber and steel piling have long been used as a means of increasing the shear strength of a failing mass by multiple driving of units into a slide area. While this has been successful in many applications, where the slide materials contained rock slabs or boulders or are underlain by bedrock, the piling were rarely able to penetrate sufficiently below the failure surface to achieve stability. Recent development of heavy duty pier drilling equipment and improved cutting tools has created a capability for installing reinforced concrete cantilever piers which could be conventionally designed and constructed, enabling support of the landslide forces through the passive restraint provided by the underlying competent strata. Figure 1 presents the concept used in the case histories which follow.

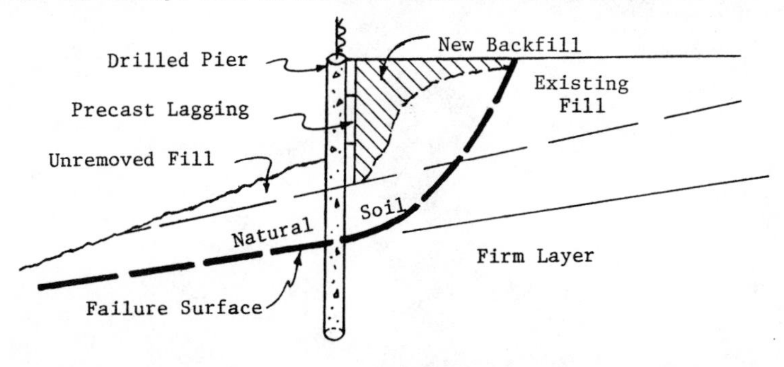

FIG. 1. Pier Wall Concept

1 Geotechnical Engineer-Vice President - The H. C. Nutting Company
Cincinnati, Ohio

Satisfactory local experience has been achieved using 18" to 30" (457 to 762 mm.) diameter piers spaced 5 to 7 ft. (1.5 to 2.1 m) on center when the depth to the failure surface is generally less than 20 ft. (6.1 m). Figure 1 presents a typical cantilever pier wall application in the correction of a roadway fill slope stability problem.

A single line of individual piers forms a continuous structure through the principle of soil arching which allows restraint of the active forces tending to promote movement of material between the piers. Shallow lagging is normally used to support the fill needed to restore the grade. Figure 2 illustrates the arching principles involved. The safe open space between piers is directly related to the applied forces, soil strength, groundwater levels and other factors. Local experience has been satisfactory when using an open space between piers of 2 to 3 pier diameters.

Longitudinal Pier Spacing

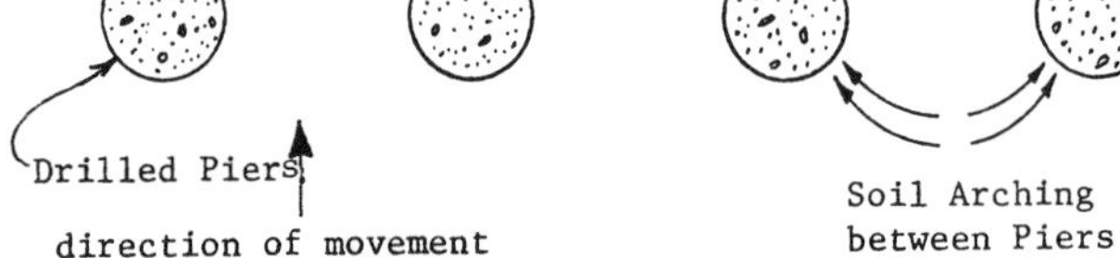

FIG. 2. Soil Arching Principle

Examples of Actual Installations

A cut was made into the toe of a colluvial slope of an ancestral pre-glacial valley which had an unbraced vertical height of approximately 15 ft. (4.6 m) and a length of 500 ft. (153 m). Several hundred feet

FIG. 3A. Tension Cracks in Roadway Paralleling Cut

upslope, cracks appeared in a roadway which paralleled the cut and contours, conveying initial signs of movement (Figure 3A). The magnitude of movement was sufficient to cause closing of the roadway. A pier wall was constructed along the berm to support the roadway (Figure 3B) consisting of 30" (762 mm) diameter piers spaced 5 ft. (1.5 m) on center, using cast-in-place concrete, cylindrical columns and lagging to create the finished grades.

FIG. 3B. Pier Wall Correction (FIG. 3A) along Roadway Berm.

FIG. 4. Pier Wall Using Recycled Reinforced Concrete Lagging.

Long-term erosion of the toe of a slope by a major river caused an acceleration in the rate of creep movement. A secondary roadway existed midway up the slope and at a bend, a sidehill fill had been constructed when creating the original alignment and grade. In time, slope movements caused tension cracks to form in the roadway and an unsafe surface for gravel trucks hauling material from a nearby aggregate pit. A drilled pier wall was used to provide support for approximately 7 ft. (2.1 m) of new fill required to restore the roadway grade (Figure 4). Eighteen inch (457 mm) diameter concrete piers were placed 7 ft. (2.1 m) on center to create the 200 ft. (61 m) long restraining structure. Used reinforced concrete railroad ties were placed as lagging for support of the granular backfill needed to create finished grade.

A sidehill fill had been placed across an erosional gully to form a secondary roadway. Long-term seepage into the natural foundation soils combined with removal of material from the toe of slope by erosion produced slope failure and loss of one-half of the roadway for a 50 ft. (15.3 m) long section. Pier wall corrections included 24" (610 mm) diameter piers with formed square columns placed 7 ft. (2.1 m) on center (Figure 5). Precast concrete crib wall sections were used for lagging.

FIG. 5. Pier Wall Using Formed Columns and Concrete Lagging

A residential roadway traversed the crest of a ridge, and portions of the alignment paralleled the edge of a rock cliff leading to a major valley. Long-term weathering of the layered shale and limestone bedrock forming the cliff caused the top of slope to encroach upon the roadway. Structural support was provided using a pier wall (Figure 6) consisting of 24" (610 mm) diameter piers with circular formed columns placed 5 ft. (1.5 m) on center. Lagging between piers was 3 ft. (0.9 m) deep precast concrete panels.

Long-term creep movements caused loss in elevation of the downhill berm of a narrow hillside roadway. Localized street improvement called for reconstruction and widening. Necessary structural support was provided using 30" (762 mm) diameter piers placed 5 ft. (1.5 m) on center (Figure 7). A 4 ft. (1.2 m) deep cast-in-place closure panel provided the support needed for gently sloping backfill and new roadway.

FIG. 6. Pier Wall Using Precast Panel Lagging.

FIG. 7. Pier Wall Using Cast-in-Place Concrete Lagging.

FIG. 8. Pier Wall Using Steel Columns and Lagging.

An unbraced cut was made at the toe of a slope in a residential area to create a level area for a new service station and parking. The cut exposed layered bedrock, fully penetrating the overlying colluvium. A slide occurred, overburden materials accumulated at the base of slope and a frame dwelling on the slope was enveloped by the scarp with resultant settlement and cracking. Corrections included underpinning of the home and construction of a pier wall at the toe of slope using 24" (610 mm) diameter piers placed 7 ft. (2.1 m) on center. Rolled steel, wide flange sections were used for reinforcing and projecting columns with steel channel sections for lagging. A 6 ft. (1.8 m) height of wall was selected to allow flattening of the backslope for stabilization grading (Figure 8).

A residence was constructed on a moderately steep slope overlooking the panoramic view of a river valley below. A sidehill fill had been placed adjacent to the downslope side of the residence to form a landscaped terrace. Long-term creep movements caused a joint to open in a sewer which extended from the residence down the slope to a trunk sewer in a parkway below. Both the scarp and toe of slide were well defined by the movement. A pier wall was found to be cost effective and was selected to provide structural restraint. The wall was positioned at the toe of slide and 6 ft. (1.8 m) of fill added to act as a counterberm and eliminate disturbance to the landscaping on the remnant terrace. Twenty-four inch (610 mm) diameter piers placed 5 ft. (1.5 m) on center were used with galvanized steel wide flange sections for reinforcing. Lagging between piers for support of the backfill was galvanized corrugated steel decking (Figure 9).

A cut was made for a roadway at the toe of a colluvial slope, causing a slide in the overburden on the sloping surface of bedrock. A pier wall was used for structural support consisting of 30" (762 mm) diameter piers placed 5 ft. (1.5 m) on center. Cast-in-place concrete was used for the 7 ft. (2.1 m) maximum height of retaining wall needed to allow flattening of the backslope and related grading (Figure 10).

FIG. 9. Pier Wall at Toe Using Steel Columns and Lagging.

FIG. 10. Pier Wall Using Cast-in-Place Concrete

Investigation

Careful and thorough site reconnaissance is made to establish the overall limits of the slide and indicated mode of failure. Notes are recorded on those conditions which relate the geologic setting and origin of site materials, historic grading, drainage and other changes accomplished by man, slide boundaries, any surface or subsurface seepage evidence, and similar items which would influence the design analysis. Preliminary measurements are taken to establish the magnitude of movement and preliminary plan and profile sketches prepared relating landslide features. Often taped measurements and hand level elevations are included on the sketches.

Previously developed site and subsurface data pertinent to the immediate slide area are retrieved from record storage and studied. Published geologic information is reviewed and available topographic drawings are analyzed relative to time-related changes, slope and drainage features. A boring program is then designed to allow definition of material properties and the failure surface, coincident with the anticipated alignment of proposed structure, with selected borings located both above and below. Conventional sampling and boring procedures are followed and the depth of exploration is generally 1.5 to 2 times the depth to the failure surface from existing grade. Often, on large projects, slope indicator casing is installed to define the failure surface and monitor long-term movement. Groundwater levels are carefully recorded as the borings are advanced and at various intervals following completion. Selected borings are often converted into piezometers through insertion of 2" PVC pipe and porous backfill after completion of the sampling.

The laboratory testing program normally includes determining the classification and strength properties of the materials, both above and below the failure surface, to establish active pressure acting on the future pier wall structure and passive pressure available to resist overturning moment.

The developed field and laboratory test data is then assembled and appropriate summary plan and profile drawings made presenting the results of the field and laboratory investigations.

Design

Various general site considerations influence and sometimes govern the geotechnical and structural design requirements and must be considered. These may include the following.

1. Limitations imposed by property line and/or right-of-way.
2. Effect on adjacent existing buildings, sewers, water lines, gas lines, services, roadways, etc.
3. Finish grade requirements.
4. Surface and subsurface drainage features.
5. Site aesthetics and selection of above grade finished structure.
6. Access limitations for construction equipment.
7. Temporary maintenance of traffic.

A drilled pier retaining wall consists of a group of individual piers subjected to lateral loading. This differs somewhat from a single laterally loaded pier or pile. Design generally requires the satisfaction of two basic criteria. These include the pier developing sufficient resistance to ultimate failure and also acceptable deflections at working loads. For the application discussed herein, there are several conditions which require special consideration, often allowing for simplification. These are:

1. Pier behavior is usually as a short cantilevered rigid pole,
2. Deflections are usually not a governing factor in design,
3. The pier is socketed into bedrock or other competent strata below the failure surface,
4. The wall consists of a group of piers in a single row perpendicular to the direction of loading.

Short rigid behavior has been typical for most cases, since the depth to the failure surface is usually less than about 20 ft. (6.1 m) and embedment lengths below the failure surface are usually short, resulting in a L/D ratio less than about 10. Ultimate capacity of the supporting material is therefore the governing factor in most designs.

Due to the placement of a row of piers at some point usually on a hillside, it is commonly the case where deflections are not important. It is expected that deflections would be relatively low for "rigid piers", especially at working loads. Deflections can get large, however, at the ultimate load and for piers which exhibit more flexible behavior. For pier wall applications where deflection is an important consideration, various design methods for estimating deflection are given in the references. (2)(3)(4).

The principle for utilizing drilled piers is to assure sufficient penetration below the failure surface and into underlying competent strata and create a close enough spacing between piers to safely resist driving forces. Since this involves penetration into material having higher shear strengths, the most reasonable approach must account for the layered system to realistically (as possible) model the earth pressure distribution.

The lateral earth pressure and its distribution must consider group effects. There is evidence that the ultimate lateral resistance of the pile group is considerably less than the sum of the ultimate lateral resistances of the individual piles in the group. This seems to be the case when pile spacing is less than about 4 pile diameters (2). This, however, is a function of many variables including pile length, diameter, spacing, group configuration, and others.

For the drilled pier wall it is reasonable to assume 2-dimensional earth pressure developing along a continuous surface. Actually, this is conservative as it neglects shear stresses acting along the sides of the pier.

Local use of the design method derived from that presented by Ivey & Hawkins (7) has been successful and has yielded what has been considered reasonable results. More sophisticated methods of analyses, however, are needed for large slide applications. The method presented herein allows for simplicity of solution by normal hand methods. The following assumptions have been made in this analysis:

1. Two dimensional, with earth pressure developing continuously along the pier wall.
2. Rankine earth pressure is used where no friction is assumed to occur at the soil-wall interface.
3. Passive earth pressure above the failure surface has been neglected on the downhill side of the piers. This accounts for long-term creep and other effects which would cause separation of the soil from the shaft and eventual loss of resistance. For certain applications, the validity of this assumption may change.
4. The applied load is static (cyclic loading effects on loss of resistance are not evaluated), and
5. A safety factor of 2 is used to compute allowable passive resistance along the entire embedment length (L).

An example calculation using this method and the simplified case is presented in the following (Figure 11).

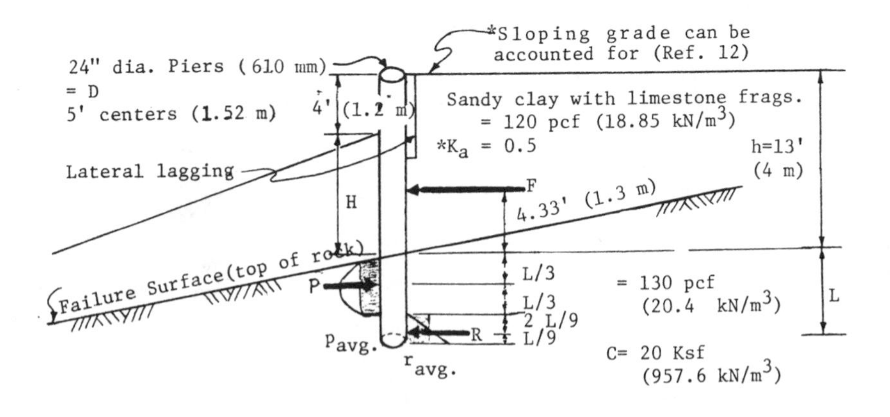

FIG. 11. Typical Calculations

$F = K_a \gamma h^2/2$ .... (1)
$= 0.5 \times .12 \times 13^2/2 = 5.07$ Kips/ft. (73.97 kN/m)
5' centers = 5 x 5.07 = 25.35 K/pier (112.76 kN)

Summation of moments about P = 0 .... (2)
which yields 25.35 (4.33+L/3) = R (5/9L)
Try L = 6.5 ft. (1.98m) from Eqn(2) R = 45.63 Kips (202.96 kN)

Summation of horizontal forces = 0 .... (3)
which yields P=F+R = 25.35 + 45.63 = 70.98 Kips/pier (315.72 kN/pier)

$p_{avg.} = \dfrac{P}{2/3\ LD}$ .... (4a)

Due to parabolic pressure distribution $p_{max.} = 1.5\ p_{avg.}$ .... (4b)

$r_{avg.} = \dfrac{R}{1/3\ LD}$ .... (5a)

Due to triangular pressure distribution $r_{max.} = 2\ r_{avg.}$ .... (5b)

The ultimate passive pressure @ P or L/3
$= \gamma_{rock}\ (L/3) + \gamma_{soil}\ (H) + 2C_{rock}$ .... (5c)

$= 0.13\ \dfrac{(6.5)}{3} + 0.12(9) + 2(20) = 41.36$ Kips/ft.^{2} (1980.32 kN/m^2)

Also from Eqn (4b) $p_{max.} = \dfrac{1.5(70.98)}{2/3(6.5)(2)} = 12.30$ Kips/ft.^{2} (588.92 kN/m^2)

SF = 41.36/12.30 = 3.4 > 2

The ultimate passive pressure @ R or 8/9L
$= \gamma_{rock}\ (8/9L) + \gamma_{soil}\ (h) + 2\ C_{rock}$ .... (5d)

which yields 0.13(8/9)6.5+0.12(13) + 2(20) = 42.31 Kips/ft.^{2} (2025.80 kN/m^2)

from Eqn (5a) $r_{avg.} = \dfrac{45.63}{1/3(6.5)(2)} = 10.54$ Kips/ft.^{2} (504.66 kN/m^2)

SF = 42.31/10.54 = 4.0 > 2

The ultimate passive pressure @ L
$= \gamma_{rock}\ (L) + \gamma_{soil}\ (h) + 2\ C_{rock}$ .... (5e)

$= 0.13(6.5) + 0.12(13) + 2(20) = 42.40$ Kips/ft.^{2} (2030.11 kN/m^2)

From Eqn (5b) $r_{max.} = 2(10.54) = 21.08$ Kips/ft.^{2} (1009.31 kN/m^2)

SF = 42.40/21.08 = 2.0 $\geq$ 2

Using the above method, the correct embedment length can be determined. The point of zero shear can then be estimated from the developed pressure distribution and the maximum bending moment obtained by the summation of moments about this point using a free body diagram. Structural reinforcing can then be conventionally designed. Rolled shapes have advantage for certain applications, particularly when using treated timber lagging or precast concrete panels. Occasionally, used steel beams can be obtained which would satisfy the moment requirements and allow economic advantage. Composite sections of deformed bar reinforcing below grade with rolled steel sections above grade are used.

Construction

Drilled pier walls are constructed using conventional drilling equipment and concrete placement procedures. Minor pre-grading is normally required to create an approximate 12 ft. (3.7 m) wide bench for access of the drilling and subsequent concrete equipment. Sites having steep terrain with limited access can utilize a track-mounted drill with subsequent concreting using pumping procedures. For the typical pier wall constructed along the berm of a roadway, conventional crane or truck-mounted pier drilling equipment is normally used.

Routine pier drilling procedures are followed to extend the pier to the depth below the failure surface required by design. Careful surface inspection of the cuttings recovered on the flights of the auger as the hole is advanced is sufficient to allow definition of the materials being penetrated, normally eliminating necessity for an in-hole inspection by project personnel. When caving of the open hole is encountered, standard casing is used.

Fabrication of deformed bar reinforcing cages is performed in an adjacent area of the site, as close as practical to the drilling. Placement of steel in the open hole utilizes a tagline from the pier drilling rig. For non-uniform cages having larger bars on the uphill or tension side, care is required to assure field placement is consistent with design requirements. Rolled steel H or wide flange sections are placed in the same manner as cages unless the weight exceeds the safe lifting capacity of the pier rig, in which case a separate crane is used.

The desired drilling sequence is to proceed from one end of the pier wall structure towards the other with steel placement and concreting completed on each pier on the day of drilling. Under certain circumstances, it may be desirable to construct every other pier followed by completion of the intermediate piers as a second stage of construction.

Where concrete piers project above grade, rectangular, square or cylindrical formed sections have been used. Following curing, forms are removed.

Backfill construction is usually accomplished in stages as the individual lagging is placed. The type of backfill and compaction requirements are established by the design. Normally, cohesionless granular materials are desired immediately behind the structure and for a lateral distance at least equal to one-half of the height projecting above grade. A drainage outlet is usually provided on the downslope side of the structure to eliminate ponding of seepage and long-term effects. Spaced aggregate trench or perforated pipe drains are used to provide a seepage outlet. A typical construction sequence is shown in Figure 12.

FIG. 12. Typical Pier Wall Construction

Cost

The cost of constructing a drilled pier retaining wall is comparable to other forms of reinforced concrete structure. The designer, however, has considerably more flexibility in his selection of the materials forming the finished structure as variations in the type and size of member have a critical effect on cost and aesthetics rarely is a governing criteria. Many applications of recycled steel and precast concrete construction materials are possible without affecting structural integrity. Precast concrete panels for lagging have a considerable advantage over cast-in-place concrete structure. Many times, cost-effective revisions are suggested by the pier drilling specialty contractors themselves.

Reinforcing steel has a significant influence on the cost of drilled pier wall structure. The design must carefully consider the resisting moment which must be provided for by the steel reinforcing and the alternate shapes of rolled steel sections or deformed bar cages which would be satisfactory. Usually, larger diameter piers will allow more effective use of the reinforcing and a lower total cost of construction.

The majority of the various items included in a landslide correction using pier wall structure can be estimated using current Highway Department bid items and unit rates. Recently completed local projects using 24" (610 mm) diameter piers placed 5 ft. (1.5 m) on center and deformed bar reinforcing have resulted in a cost of approximately $400.00 per cu. yd. for in-place concrete pier structure alone, excluding the added cost of precast concrete panel lagging, grading, paving, guardrail and similar items.

Limitations

Drilled pier walls are an effective form of restraining structure for correction or prevention of earth slope stability problems when utilized according to the principles which have been described. They have, however, inherent limitations which must be considered when evaluating their possible use.

For fill slope stability applications, the normal tendency is to place the wall at the crest of the slope, nearest the roadway or other improvement which it is intended to support. Inherently, only a small portion of the force causing the landslide is supported by new structure. A continuation of downslope movement then occurs below the pier wall causing a greater depth of the piers to become exposed. When sufficient movement has occurred to expose the bottom of lagging between piers, loss of granular backfill material soon follows. Where a vertical face is created by the arching action of cohesive materials between piers, long-term effects of weathering (wet/dry, freeze-thaw) cause loss of strength and removal of material from between and behind adjacent piers. Figure 13 illustrates such an occurrence for a single line of unconnected individual drilled piers placed along the berm of a roadway.

FIG. 13. Continuation of Slope Movement below Wall.

Continuation of movement of the materials downslope from a pier wall may be accompanied by a renewal of old distress or allegations of new distress in structures on adjacent property at the toe. This can trigger litigation with accusation that the movement was caused by the new drilled pier structure. Any removal of material from the toe of slope by the adjacent owner, which many times was the initial cause of the instability, only serves to accelerate movement of the materials below the pier wall. To offset this inherent limitation and increase support of the forces causing movement, it is desirable that the pier wall be placed as far down the slope as possible. It is equally important to eliminate the placement of any new fill downslope from the wall, including the spoil materials removed from the drilling operations. This is why lagging is used between piers for support of the backfill needed to create finished grade. The backslope of the materials placed uphill from the pier structure must also be carefully chosen to prevent overtopping.

Local experience has demonstrated a tendency for enlargement of the flanks of a slide several years following completion of corrections by pier wall structure. Where this has occurred, it is usually the result of a continuation of stream erosion or additional excavation at the toe of slope which caused the original instability or changes in the flow of surface or subsurface seepage. It is therefore normally advisable to extend the pier wall several sections past the flank of a slide and into adjacent stable material.

Economic limitations govern the use of cantlievered piers, as the force increases as a function of the square of the height. This results in an increase in the required depth of penetration below the failure surface, a greater diameter of pier and additional reinforcing to resist the overturning moment. Alternate use of anchor tie-backs should, therefore, be investigated on projects where their installation is practical.

Acknowledgement

The writer wishes to acknowledge and express appreciation to my former colleague, Jess A. Schroeder, who is now a Project Engineer with Soil Testing Engineers Inc., for his assistance in preparation of the design considerations and analytical procedures which are presented.

Appendix I.--References

1. Bhushan, K., Haley, S. C., and Fong, P. T., "Lateral Load Tests on Drilled Piers in Stiff Clays," *Journal of the Geotechnical Engineering Division,* ASCE, Vol. 105, GT8, Proc. Paper 14789, Aug., 1979, pp. 969-985.
2. Broms, B. B., "Lateral Resistance of Piles in Cohesive Soil," *Journal of the Soil Mechanics and Foundations Division,* ASCE, Vol. 90, SM2, Proc. Paper 3825, Mar., 1964, pp. 27-63.
3. Broms, B. B., "Lateral Resistance of Piles in Cohesive Soils," *Journal of the Soil Mechanics and Foundations Division,* ASCE, Vol. 90, SM3, Proc. Paper 3909, May, 1964, pp. 123-156.
4. Broms, B. B., "Design of Laterally Loaded Piles," *Journal of the Soil Mechanics and Foundations Division,* ASCE, Vol. 91, SM3, Proc. Paper 4342, May, 1965, pp. 79-99.
5. Davisson, M. T., and Salley, J. R., "Lateral Load Tests on Drilled Piers," *Performance of Deep Foundations,* ASTM STP 444, American Society for Testing and Materials, 1969, pp. 68-83.

6. Davisson, M. T., and Prakash, S., "A Review of Soil-Pole Behavior," *Highway Research Record*, No. 39, Nov., 1963, pp. 25-48.
7. Ivey, D. L., and Hawkins, L., "Signboard Footings to Resist Wind Loads," *Civil Engineering*, Dec., 1969, pp. 34-35.
8. Kocsis, P., Discussion of Paper by Bhushan, K., Haley, S. C., and Fong, P. T., *Journal of the Geotechnical Engineering Division*, ASCE, Vol. 106, GT 10, Oct., 1980, pp. 1172-1174.
9. Poulos, H. G., and Davis, E. H., *Pile Foundation Analysis and Design*, John Wiley and Sons, New York, 1980.
10. Reese, L. C., and Allen, J. D., *Drilled Shaft Design and Construction Manual*, Vol. Two - Structural Analysis and Design for Lateral Loading, U. S. Dept. of Transportation, Federal Highway Administration, Offices of Research and Development, Implementation Division, Washington, D.C., July, 1977.
11. Reese, L. C., and Matlock, H., "Non-Dimensional Solutions for Laterally Loaded Piles with Soil Modulus Assumed Proportional for Depth," *Proceedings*, Eighth Texas Conference on Soil Mechanics and Foundation Engineering, Austin, Texas, 1956.
12. *USS Steel Sheet Piling Design Manual*, July, 1975.
13. Welch, R. C., and Reese, L. C., "Lateral Load Behavior of Drilled Shafts," *Research Report 89-10*, Cooperative Highway Research Program with the Texas Highway Department and U. S. Department of Transportation, Federal Highway Administration, by The Center for Highway Research, the University of Texas at Austin, May, 1972.
14. Woodward, R. J., Jr., Gardner, W. S., and Greer, D. M., *Drilled Pier Foundations*, McGraw-Hill Book Company, New York, 1972.

Appendix II.--Notation

The following symbols are used in this paper:

C = cohesion,
D = pier diameter,
F = active earth pressure resultant,
K_a = active earth pressure coefficient,
L = pier embedment length (in supporting stratum),
H, h = depth to failure surface,
p = soil reaction,
$p_{avg.}$ = average reaction pressure in upper 2/3 of L,
$p_{max.}$ = maximum reaction pressure in upper 2/3 of L,
P = resultant reaction force in upper 2/3 of L,
$r_{avg.}$ = average reaction pressure in lower 1/3 of L,
$r_{max.}$ = maximum reaction pressure in lower 1/3 of L,
R = resultant reaction force in lower 1/3 of L,
γ = unit weight of soil or rock
SF = safety factor

SLIDE STABILIZATION AT THE GEYSERS POWER PLANT

By H. John Hovland,[1] M.ASCE and Donald F. Willoughby,[2] A.M.ASCE

INTRODUCTION

The Geysers Geothermal Power Plant is located approximately 100 miles (161 km) north of San Francisco, California. This power plant is operated by the Pacific Gas and Electric Company. The Power Plant consists of individual units which use geothermal steam collected from wells in surrounding leaseholds to generate electricity.

The Geysers area is characterized by very steep, rugged terrain. Much of the area is in a prominent geologic assemblage known as the Franciscan Formation. This formation contains a large percentage of rock which has been chemically altered, faulted, or sheared. The numerous weak, altered, or sheared rocks in the Franciscan Formation, the steep slopes, and heavy rainfall (the area generally receives no rainfall in the summer, but up to 100 inches (254 cm) of rain may fall at some locations during the fall and winter) have caused numerous active landslides and dormant landslide deposits.

Slope stability is a very important consideration in the siting of power plant facilities at The Geysers. However, a power plant unit must be within 1-2 miles (1.6 to 3.2 km) of all wells which supply steam to it in order to make effective use of the geothermal steam resource. These constraints sometimes require innovative and unusual solutions to slope stability and foundation engineering problems.

REVIEW OF LITERATURE

Slide stabilization with a retaining wall tied back with rock anchors is a rather challenging lateral earth pressure problem. Because of possible movements within the slide, and restraint at the top of the wall by the tie-back anchors, a limit equilibrium state of minimum active earth pressure cannot be assumed. Deformations of the wall are likely to be somewhere between earth pressure at rest and minimum active earth pressure conditions. The general nature of strains required to produce such intermediate as well as limiting

---

[1] Sr. Civil Engineer, Pacific Gas and Electric Co., San Francisco, California
[2] Civil Engineer, Pacific Gas and Electric Co., San Francisco, California

earth pressure conditions is discussed by Henkel (7). According to the classical experiments by Terzaghi (15), for the case of horizontal back fill, the top of a wall retaining dry sand must move out approximately 0.0005H (H = height of wall) to develop the minimum active case. According to Morgenstern and Eisenstein (10), where deformations can have a significant impact on lateral pressures or their distribution, semi-empirical methods are used. Gould (6), who reports on some significant case histories, concludes that the estimation of earth pressures for complicated situations is likely to remain a semi-empirical procedure.

The use of large diameter, cast in-place, reinforced concrete piers (also called caissons or piles) has been described by Shannon & Wilson, Inc. (14), DeBeer and Wallays (4), Andrews and Klasell (2), Anderson (1), and Merriam (9). Where the application of such piers has been based on rational design, they have performed well as components of a retaining wall or in slope stabilization.

Reti (13) describes an anchored retaining wall, 20 ft (6 m) high with a back-slope 5 to 15 ft (1.5 to 4.5 m) high. The facing consisted of 8-in. (20 cm) reinforced gunite between concrete pilasters. Two 100-ft (30.5 m) long tie rods anchored each pilaster to unweathered sedimentaries. For the last 60 ft (18.3 m), 13 psi (89.7 kN/m^2) bond was assumed. The anchors were tested to 150 kips (668 kN), which resulted in a 2.5 in. (6.35 cm) elongation, and the wall moved back into the hillside about 3/4 in (1.9 cm). The anchors were locked at 100 kips (445 kN). K = 0.8, more than twice K_a. Another caisson-tieback retaining wall is described in the Engineering News Record (5). The wall was constructed at the toe of a 300-ft (91 m) high hill in San Francisco, with a slope of about 1:1.5. The wall is 17 ft (5.2 m) high and 680 (207 m) ft long, with 30-in. (76 cm) diameter caissons, 23 ft (7 m) long, spaced at 8 ft (2.44 m) center-to-center. One-ft (30.5 cm) diameter friction tie-backs, which are 32 ft (9.8 m) long, retain the top of the caissons. Facing consists of 5.5 in. (14 cm) thick, 4,000 psi (27,600 kN/m^2) pneumatically applied concrete.

There have been relatively few applications of a retaining structure consisting of piers and tie-back anchors. The last two case histories described above have certain similarities to The Geysers Unit 14 retaining wall to be described subsequently. However, there are also significant differences, and because of these differences and the degree to which design of such facilities depends on full scale case histories, The Geysers 14 retaining wall may constitute an important addition to our state of knowledge.

THE LANDSLIDE

The Geysers Unit 14 is situated in the southern portion of The Geysers Geothermal Area, on a narrow northeast trending spur ridge. Bedrock underlying the northeast end of the site is relatively massive serpentinite. An active landslide originally covered some of the serpentinite. This landslide was removed during construction, and a turbine-generator building was constructed on the serpentinite bedrock. The southern portion of the site, the location of the cooling tower, is underlain by graywacke/schist bedrock. A thick blanket of colluvium covered the graywacke/schist bedrock. The slope of the colluvium varied, but it was steeper than 1½:1 in some areas above and below the power plant site.

As originally planned, the slope was to be excavated at 1½:1 with intermediate benches. The cut slope was to begin 125 ft (38 m) above the plant site. A retaining wall, up to 30 ft (9 m) high, was to be constructed at the toe of the slope. Rockbolts were to be installed through the wall and anchored into the graywacke/schist bedrock to resist high sliding forces.

Excavation of the power plant site began in the spring of 1977. When the excavation of the cut slope was 50 ft (15 m) above the proposed plant grade, the cut slope was steepened to 3/4:1 to facilitate construction of the retaining wall. This excavation was approximately 25 ft (7.6 m) above plant grade when a landslide occurred in the southernmost portion of the cut slope. A plan of the landslide is shown in Fig. 1. An earth buttress was immediately placed at the toe of the landslide to prevent further movement.

The stability of the cut slope had been analyzed before construction for an excavation case which corresponded to the slope configuration when the landslide occurred. Two exploratory borings had been drilled into the colluvial slope during the siting phase. Triaxial, consolidated, undrained strength tests were performed on samples of a clayey soil recovered from the colluvium/bedrock interface. The total stress strength parameters used were $\phi=19.5^{\circ}$ and c=800 psf (38.3 kN/m^2) at 5% strain. The materials exhibited strength increase with strain, and at 20% strain, c=1250 psf (59.9 kN/m^2); ϕ did not change. The effective stress strength values were $\phi'=35^{\circ}$, c=0. The total stress analysis was found to be the critical case. The pre-excavation factor of safety was calculated to be 1.47. The factor of safety for the excavation case was calculated to be 1.41, a 4% reduction.

Fig. 2 is a section through the landslide. As shown, and as also determined from field observations, most of the sliding plane was in colluvium. Only a small, deeper part of the sliding plane followed the clayey colluvium/bedrock interface. The section in Fig. 2 and others through the landslide were used to back-calculate the strength of the colluvium when the landslide occurred. Because of the granular nature of the soil involved in the landslide, c=0 was assumed. The back-calculated total stress strength was found to be $\phi=24^{\circ}$.

While it is difficult to pinpoint a single most important cause of the landslide, there are significant contributing factors. It appears that the measured soil strength properties were not representative of the materials involved in the slide. The clayey soil at the colluvium/bedrock interface was the weakest soil encountered during pre-slide explorations. However, most of the sliding plane was within the colluvium, which consists of brown sandy silt and semi-angular, mostly 2 in. (5 cm) fragments of weathered greenstone, with a clayey or silty coating; the voids between the greenstone fragments were partially filled with similar fines. At the time of the pre-slide explorations, the colluvium was relatively dry.

In fact, no water table was found in the exploratory borings. However, as the excavation for the site aproached the elevation at which the slide occurred, some seepage was noticed coming out of the hillside. Subsequently, large quantities of seepage water have been continually drained out of the hillside through approximately 3700 ft (1130m) of horizontal drains. Similar conditions have been encountered elsewhere at The Geysers - significant seepage after a major excavation even close to a ridge top. It appears that elastic heave caused by the removal of the excavated mass opens fractures within the rock, which permits a rearrangement of the ground water regime.

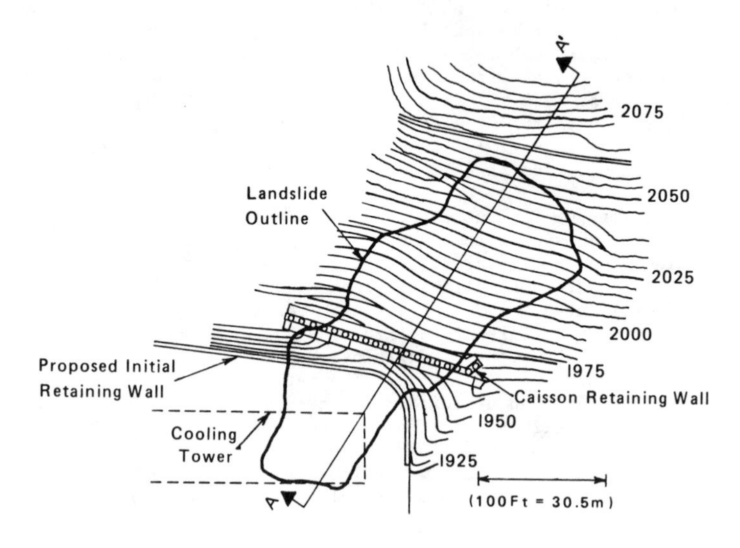

Fig. I - Plan of Landslide Outline, Retaining Walls, and Final Grading

Other slope failures have been reported where the pre-slide factor of safety was significantly greater than 1. Pilot (11) found that for 5 cases for which the factors of safety were from 1.23 to 2.06, the failures could be attributed to poor definition of strength, unrepresentative samples, and evolution of excess pore pressure.

STABILIZATION

Various alternatives were considered for stabilizing the slope. These included a caisson-tieback wall, crib and bin walls, and crib and bin walls at more than one elevation. The alternatives fell into two broad categories: (1) walls which require the preparation of a competent foundation by means of an excavation, (2) walls where the competent foundation support is found without an excavation. A larger excavation to find competent foundation conditions would have probably been a futile effort; the slide would have progressed deeper and

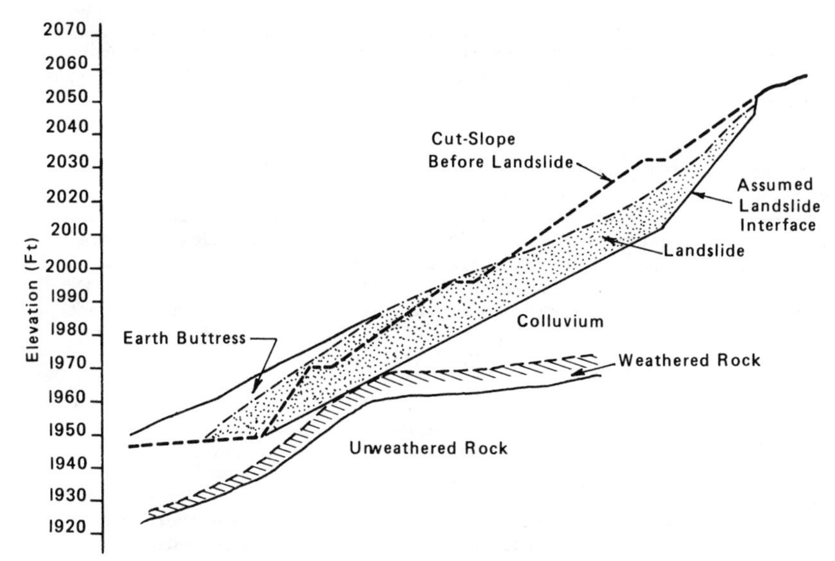

Fig. 2 - Section A-A' Through Landslide

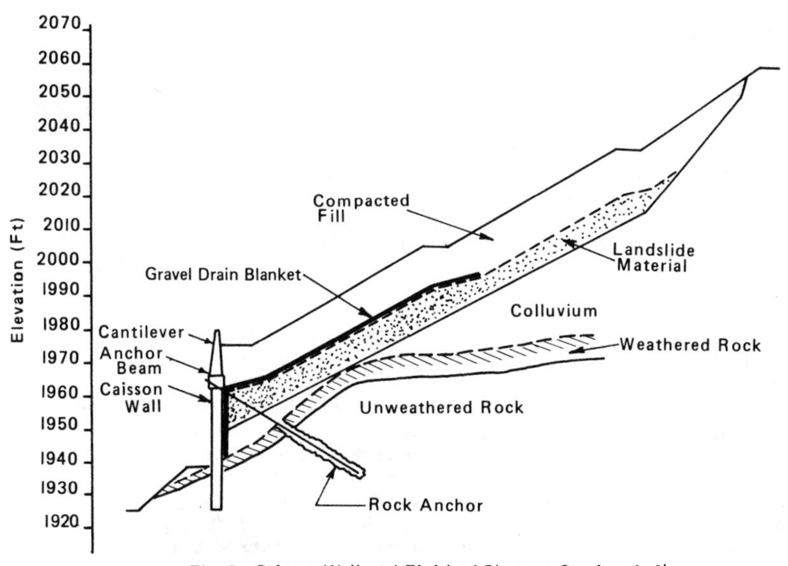

Fig. 3 - Caisson Wall and Finished Slope at Section A-A'

further upslope. Therefore, conditions pointed to a retaining structure which could be constructed through the landslide, and where competent support could be found without an excavation. A caisson-tieback retaining wall, as illustrated in Fig. 3, was selected.

The conditions at Geysers 14 were unconventional in several important ways. The structure had to retain soil and landslide materials, but be tied to rock by piers and rock anchors. Therefore, the loads were determined from principles of soil mechanics, but the support conditions were established by assuming conservative failure mechanisms in the supporting rock. In addition, sloping ground conditions exist below the proposed wall and for several hundred feet above the proposed wall, making much of the previous experience with tie-back excavations in soil inapplicable.

The retaining structure was analyzed for several assumed conditions; the most important conditions are illustrated in Figs. 4, 5, and 6. Common for all these diagrams are the wall, the toe buttress with weight W_b, and the upper portion of the earth pressure diagram with total force F_s. The wall consists of an anchor beam with tieback rock anchors, a cantilever extension above the anchor beam, and 4-ft (1.22m) diameter piers below the anchor beam to a depth of 49 ft (15m) below the ground surface on the upslope side of the wall, as shown in Figs. 4, 5, and 6. The dimensions shown are averages of as-built dimensions for ten piers in the central maximum section.

The toe buttress wedge with weight W_b is composed mostly of unweathered, blue-grey schist. The rock can be described as massive to blocky and seamy. Two to four foot blocks or layers of hard schist are separated by fresh fractures and joints, and zones of fissile, slickensided, and possibly sheared rock. Each caisson boring was carefully logged and special attention was given to through-going, unfavorably oriented rock weaknesses. No such weaknesses were found. However, the presence of the joints and rock weaknesses dictated that failure through solid rock could not be assumed. With the sloping ground surface and rock conditions as described, a potential failure plane was assumed at the base of the wedge, as shown in Figs. 4, 5, and 6. Resistance to sliding, F_b, along this assumed potential failure plane was computed using only the weight of the wedge and an internal friction angle of 30 degrees.

The upper portion of the pressure diagram is also common for conditions analyzed. The soil force, F_s, was computed from an equivalent fluid pressure of 98 pcf (15.6 kN/m^3). The upper approximately 60% of the 27 ft (8.2 m) of soil behind the retaining wall consists of compacted engineered fill. The lower approximately 40% of this 27 ft (8.2 m) of soil consists of landslide debris. The slope would strain toward the wall along the shear zone of landslide debris. Therefore, the equivalent fluid pressure of 98 pcf (15.6 kN/m^3) was assumed to be applicable to the entire 27 ft (8.2 m) of soil. However, the force so computed has been distributed in a trapezoidal fashion with the upper end pressure coming from a cohesion of 740 (35 kN/m^2) psf. The equivalent fluid pressure of 98 pcf (15.6 kN/m^3) was selected as a condition which envelops forces computed

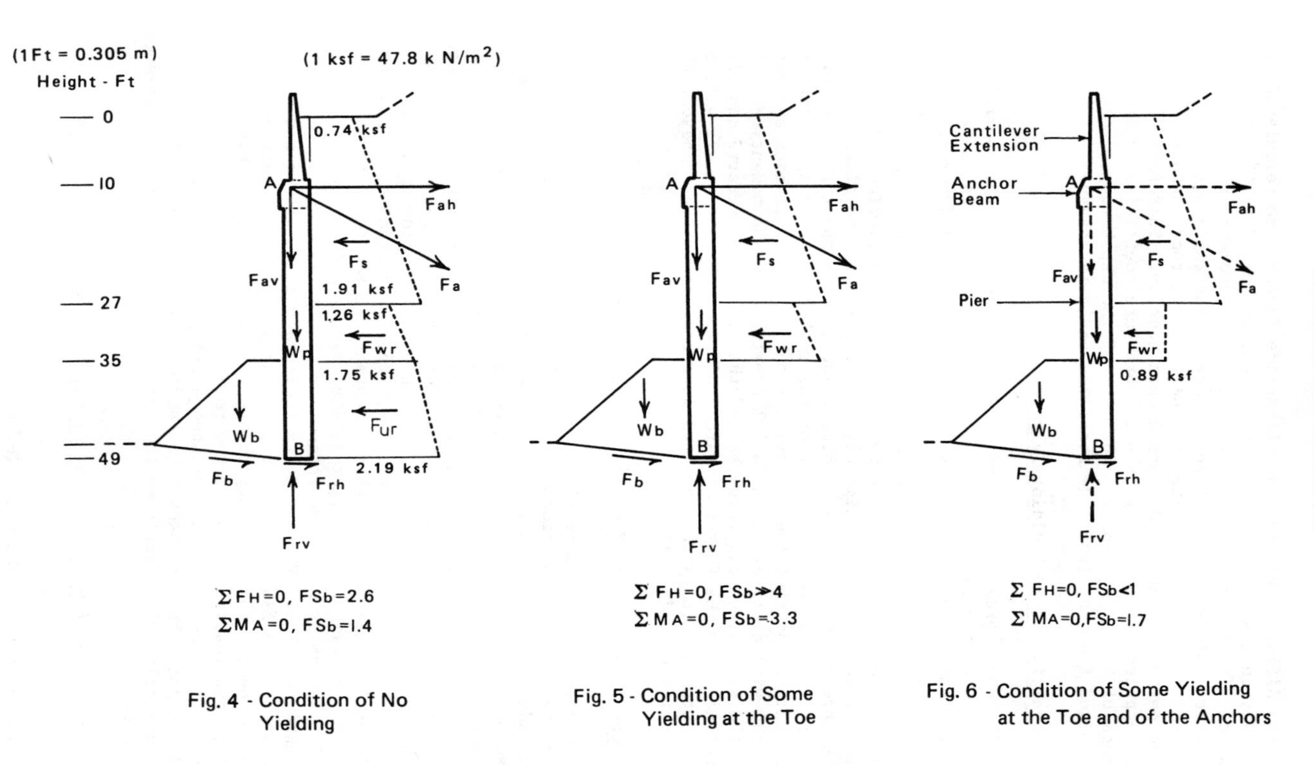

Fig. 4 - Condition of No Yielding

Fig. 5 - Condition of Some Yielding at the Toe

Fig. 6 - Condition of Some Yielding at the Toe and of the Anchors

from two different procedures. Landslide forces were calculated from stability analyses using soil properties, c=0 and $\phi = 24^{o}$, as back-calculated from pre- and post-landslide geometrical conditions of the slope. Since the structure will be rather rigid with a relatively fixed toe support and retained at the top with rock anchors, minimun active earth pressure conditions cannot develop. Therefore, forces were computed also by using an earth pressure coefficient which is the average of K_a and $K_o=1-\sin\phi$. For most of the depth range, forces determined from the landslide stability analysis governed.

Condition of No Yielding (Figure 4)

The earth pressure diagram in Fig. 4 represents the condition of conservative stability with no yielding. An at-rest earth pressure is assumed behind the piers in the zone of weathered and unweathered rock. As shown in Fig. 4, for these conditions, horizontal force equilibrium indicates a factor of safety (FS) of approximately 2.6 against sliding of the toe buttress. Moment equilibrium indicates a factor of safety of approximately 1.4 with respect to sliding of the toe buttress. Overturning or moment equilibrium about the base of the wall indicates a loading factor of the rock anchors of approximately 1.1. This means that the anchors are pulling the wall back against the soil with a moment which is 10% greater than the moment resulting from the estimated soil force behind the wall. The reserve capacity as reflected by the factor of safety of the anchors (Fig. 7) will be called upon should earth pressures increase significantly beyond those anticipated by this analysis.

Condition of Some Yielding at the Toe (Figure 5)

Should there be some yielding of the toe buttress, the piers would move out a proportional distance. It is assumed that the soil, landslide, and weathered rock forces would remain the same, since these materials would adjust with permanent and plastic deformation to the movement of the wall. The unweathered rock behind the lower portion of the piers, however, would not follow, and the lateral earth pressure would vanish. These conditions are illustrated in Fig. 5, which also shows the resulting factors of safety from equilibrium considerations. The factor of safety resulting from equilibrium of horizontal forces for available resistance at the toe buttress is much greater than 4. The factor of safety, with respect to moment equilibrium about A, for resistance at the toe buttress is approximately 3.3. These results indicate that the lower end of the wall is solidly fixed, and that yielding at the toe is very unlikely. Considering the vulnerability to overturning, taking moments about the base B, the loading factor of the rock anchors is found to be 1.3. That is, for these somewhat reduced earth pressures, the anchor force tends to pull the wall back into the soil with a moment which is approximately 30% larger then the moment resulting from soil pressures on the wall.

Condition of Some Yielding at the Toe and of the Anchors (Figure 6)

The conditions shown in Fig. 6 represent movement of the

retaining wall such that (1) earth pressures in the unweathered rock are reduced to zero; (2) earth pressures in the weathered rock behind the wall are also significantly reduced, and for this case computed using an active earth pressure coefficient K_a; (3) the rock anchors have yielded to the point that the vertical component of the anchor force F_{av}, cannot be depended on to contribute to sliding resistance, F_{rh}; and (4) the soil and landslide forces are the same as before, since these materials are assumed to follow the wall as the wall deflects. Equilibrium computations for these assumed conditions are also shown in Fig. 6. Horizontal force equilibrium gives a factor of safety of less than 1 with respect to sliding of the rock buttress; that is, without the rock anchors, the retaining structure would not be stable. The need for the rock anchors is, however, such that if the rock anchors can provide a resistance equal to 34% of their actual as-built load, horizontal equilibrium conditions would be satisfied. Moment equilibrium about A gives a factor of safety equal to approximately 1.7 with respect to sliding of the rock buttress.

Seismic Conditions and Overall Slope Stability

The geologic and seismic conditions at The Geysers have been quite thoroughly investigated. A special study, which references the most important works and summarizes our state of knowledge was performed by Bolt and Oakeshott (3). Additional studies are also in progress. Generally, the Geysers area has a higher frequency of microearthquakes, but a lower frequency of larger earthquakes of engineering significance than most of northern California. Probabilistic studies of the data base have also been performed, and such studies indicate that there is approximately a 10% probability for .2g peak rock accelerations to be exceeded within the design life of a typical geothermal unit at The Geysers. As shown in Figure 5, the static factor of safety is approximately 3.3 against translation for K=0.75 and a backfill sloping at about 30 degrees. Comparing these conditions with criteria presented by Seed and Whitman (13) indicates adequate seismic stability for the Geysers 14 retaining wall.

Various other stability analyses were made, which included areas of the slope above and below the retaining wall. The results of all such analyses are presented in Fig. 7. For the slope above the wall, the analyses indicated that the slide materials had to be recompacted to 95% of California Test Method 216-G to a depth of 13 to 15 ft (4 to 4.6 m). Total stress analysis, with c=740 psf (35.4 kN/m^2) and $\phi=19.5^o$, gave a factor of safety of 1.37. Effective stress analyses, with c'=0 and $\phi'=38.7^o$, gave a factor of safety of 1.31. A pseudostatic seismic coefficient (horizontal seismic force/ weight), k_s in Fig. 7, of 0.15 would bring this factor of safety to unity.

An effective acceleration, which would represent the time history by a seismic coefficient for pseudostatic stability analysis, is often taken as 2/3 of the peak acceleration. By this reasoning, a pseudostatic seismic coefficient of 0.15 represents seismicity with a 10% probability of exceedence in the design life of the plant. Larger accelerations can, of course, occur, but their exceedence probability would be even less than 10%. However, it is recognized that much of

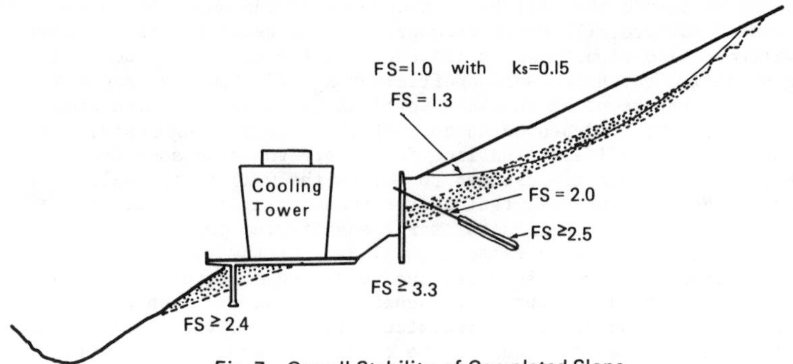

Fig. 7 - Overall Stability of Completed Slope

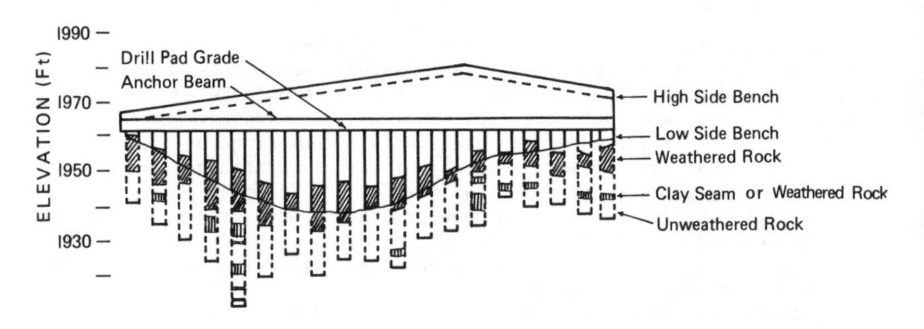

Fig. 8 - Uphill Elevation View of Caisson Retaining Wall

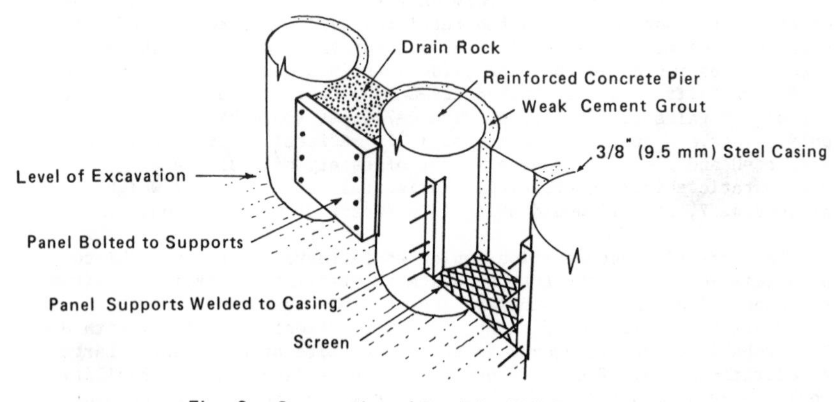

Fig. 9 - Construction of Retaining Wall Facing

the surrounding areas above the retaining wall consist of active or dormant landslides, and that the stability of the slope immediately above the retaining wall may be marginal. Therefore, some slumping or sliding in adjacent areas or above the wall should not come as a surprise either as a result of seismic conditions or due to adverse static conditions in an extremely wet year. As shown in Table 1, the conservatism of design as reflected by the factor of safety should be proportional to the consequences of failure.

Table 1

Margin of Safety vs. Consequence of Failure

Condition	Factor of Safety	Consequences of Failure
Failure of the buttress:kick-out at the toe of piers.	3.3	Would probably require shutting down plant, and reconstruction of retaining structure and slope
Failure of tie-back rock anchors	2.0	Would require installation of additional rock anchors
Slope instability	1.3	Would require stabilization and regrading above wall

CONSTRUCTION OF PIERS AND WALL

The retaining wall was constructed from the top down. (Another retaining wall constructed in this way has been described in Engineering News Record (5).) A horizontal bench was first made on the landslide surface; this was to be the elevation of the top of the piers and the location of the anchor beam, as shown in Fig. 8. Borings for the piers were drilled with a 54 in (1.4 m). diameter bucket auger. Each boring was carefully logged. Special emphasis was placed on landslide materials, the water level, seepage, colluvium, sliding planes, weathered rock, unweathered rock, and rock weaknesses. The encountered conditions are illustrated in Fig. 8. As previously described, rock weaknesses were encountered in the form of randomly oriented joints and seams. The rock is a bluish-grey schist consisting of 2 to 4 ft (0.6 to 1.2 m) blocks or layers of hard unweathered rock, separated by joints and seams. Most of the joints are clean fractures, but some of the joints and seams contain slickensided and fissile, weaker rock. Some clay seams were also encountered.

Steel casing, 48 in. (1.2 m) in diameter with a 3/8 in. (0.95 cm) wall thickness, were lowered into the borings. The reinforcing cage was lowered into the casing, and the concrete was poured. The space between the outside of the casing and the ground was filled with a weak cement grout. (Subsequently, the grout on the outside face of the piers was scraped off by hand tools to construct the retaining wall facing.) The borings were drilled and the piers installed into widely spaced borings so that only 2 or 3 widely spaced borings would be open at any one time. This was to prevent the landslide materials from moving and closing the borings. That problem was not encountered.

The anchor beam was formed and poured with sleeves cast into it for subsequent installation of the tie-back rock anchors. A cantilever wall extension was constructed above the anchor beam, since analysis indicated that it would be necessary to raise the ground upslope of the wall to provide a flatter slope. At its maximum height this cantilever extension is approximately 13 ft (4 m) high.

Temporary buttress fill and landslide materials downslope of the caissons were removed in approximately 5 ft (1.5 m) layers. The 5 ft (1.5 m) facing panels were bolted to the piers. A horizontal screen was installed at the bottom of each panel between the piers, and the approximately 1.5 (0.5 m) ft deep space between the back of the facing and the soil was filled with free draining rock resting on the horizontal screen. This process, illustrated in Fig. 9, was repeated until the excavation encountered sound rock.

INSTALLATION OF TIE-BACK ROCK ANCHORS

The installation of rock anchors started approximately at the time the cantilever extension was being built. A typical Geysers 14 rock anchor is illustrated in Fig. 10. The numbers in parentheses within this paragraph correspond to the numbers in Fig. 10. Borings (1) were drilled for the tieback anchors through the 6 in. (15 cm) diameter sleeves in the anchor beam. These borings were angled at 30° to the horizontal and extended from 44 to 95 ft (13 to 29 m) into the ground, the total length depending on the desired anchor capacity and subsurface conditions. The high-strength, steel tendons (2) were installed. Spacers kept the tendons neatly arranged and in the center of the bore hole. Primary grout (3) was injected to the desired depth for required bond strength. Next, a galvanized 4 in. (10 cm) duct (4) was placed. A bearing plate (5) and the anchor head (6) were secured, the anchor was tested to 150% of the design load, and the design load was set. Secondary grout (7) was injected through the galvanized duct so that it would squeeze to desired depth on the outside of the duct, as shown in Fig. 10. The purpose of the secondary grout is to form a tight protective seal and setting for the tendons and the galvanized duct close to the interface with the primary grout. The grease cap (8) was placed. The grease cap was necessary to protect the extended tendons, which are required for future liftoff tests. For this type of a permanent design, liftoff tests are believed essential to verify the future performance of the anchors.

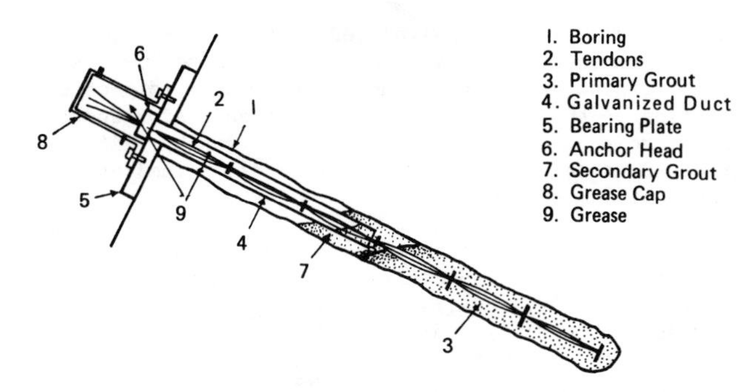

Fig. 10 - Sketch of a Geysers 14 Rock Anchor

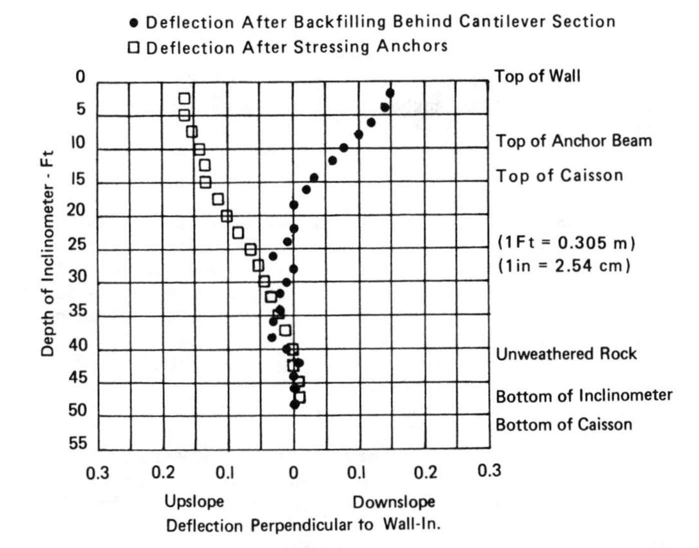

Fig. 11 - Caisson Wall Deflection

Grease (9) was injected into the galvanized duct and outside of the galvanized duct above the secondary grout level; grease was also injected into the grease cap. The grease is necessary to protect the free length portion of the steel tendons against corrosion and to allow the tendons to contract and extend freely. Finally, liftoff tests are performed as called for by the design or as indicated by future instrumentation measurements.

The tie-back rock anchors at Geysers 14 are made up of 7, 15, and 19 strands each for total capacities of approximately 145, 310, and 390 kips (645, and 1380, and 1736 kN/m^2) respectively. A bond strength of 100 psi (690 kN/m^2) was assumed in the schist bedrock. This would compare with a weak to medium rock according to Littlejohn and Bruce (8). The as-built bond stress in the anchors varies from about 25 to 40 psi (173 to 276 kN/m^2).

INSTRUMENTATION

Instrumentation and measurement of field performance is considered essential for this type of a retaining structure, because the problem is basically indeterminate, and there are great uncertainties both in the assessment of strength and deformation behavior. An inclinometer casing was inserted into one of the caissons in the center section of the wall. This casing was continued up through the cantilever wall section on top of the caissons. The inclinometer only monitored tilting of the maximum wall section. Therefore, a precise optical alignment survey was also used to monitor overall lateral movement of the wall.

Four rock anchors were instrumented with load cells to verify the actual loading in the anchors. The load cells were distributed along the wall so that a change in load on one anchor would cause a change in load on at least one load cell because of deflection of the stiff anchor beam.

Backfill was compacted behind the upper cantilever wall section before the rock anchors were stressed. The inclinometer readings for this period showed that the top of the wall moved outward 0.15 in. (0.38 cm) This tilting began at the top of the caisson, as would be expected, and as illustrated in Fig. 11. After the rock anchors had been stressed, inclinometer readings indicated that the top of the wall had moved into the slope 0.15 in. (0.38cm) from the original position; see Fig. 11. In this case, tilting of the wall began approximately where the caisson penetrated hard rock.

The first readings of the load cells were taken during the initial anchor stressing operations. The load cells provided additional assurance that the proper loads were being placed on the anchors. The pressure gage on the stressing jack apparently went out of calibration during the anchor stressing operation. This was revealed by one of the load cell readings, and a new pressure gage was installed on the jack. Without the load cells, there would not have been any check on the anchor loads and, in this case, some of the anchors would have been stressed to much less than the required design loads.

The initial readings on the load cells also provided an indication of the stiffness of the anchor beam. The loads in the cells which had been stressed first and second dropped 23% and 18%, respectively, by the end of the stressing operation. The third and fourth load cells, which were stressed near the end of the operation, showed no significant change. The stiff anchor beam deflected into the slope as subsequent anchors were stressed, thus relaxing the anchors at the first and second load cells.

The value of the redundant instrumentation became very apparent during the summer of 1978. The readings in three of the load cells began to rise at an alarming rate. However, the inclinometer and optical survey did not show movement which would justify the increased loads. The load cell readings finally peaked in October and November 1978 at levels as much as 10 times the initial lock-off loads, well above the capacity of the anchors. Then the readings declined until one load cell actually indicated a negative load. The erratic behavior may have been caused by the sun, which melted the grease within the grease caps and perhaps damaged the strain gages. The three defective load cells were replaced in 1979 when all of the anchors were re-stressed. Since that time, all of the load cells have given consistent readings. Without the reassurance of the relatively unsophisticated but reliable optical survey and inclinometer, drastic and expensive action might have been taken to ensure the stability of the wall.

CONCLUSIONS

The experiences to date with The Geysers 14 slope and retaining wall suggest the following conclusions:

1. Samples selected for testing must be representative of critical materials within the slope. It is important to be able to visualize critical conditions and materials from subtle geological features, appropriate explorations, testing, and analysis.

2. The landslide has been successfully stabilized using piers drilled through the landslide to unweathered schist, with tie-back rock anchors holding back the top of the piers and the wall.

3. Simple and redundant instrumentation proved to be indispensable during the construction and early performance of the retaining wall. Without the optical survey and the inclinometer, which provided a check on the load cells, drastic and expensive, but unnecessary action might have been taken.

4. The instrumentation will also be essential in assessing the long term behavior of the wall and the permanency of the rock anchors. For example, should the inclinometer indicate excessive down-slope movement at the top of the wall, liftoff tests can be performed to check on the load carrying capacity of the anchors.

Additional rock anchors can also be installed, should this become necessary.

APPENDIX 1. - REFERENCES

1. Anderson, Jr. A. H., "Earth Retention by Drilled Reinforced Concrete Caissons," Civil Engineering, ASCE, pp 54-56, Oct., 1976.
2. Andrews, G. H. and Klasell, J. A., "Cylinder Pile Retaining Wall," Highway Research Record, No. 56, pp 83-97, 1964.
3. Bolt, B. A. and Oakeshott, G. B., "Evaluation of Seismicity and Seismic Intensity at the Geysers Area, California," Report for P.G.& E, San Francisco, August 1978.
4. De Beer, E. E. and Wallays, M., "Stabilization of a Slope in Schist by Means of Bored Piles Reinforced with Steel Beams," Proceedings, 2nd. Congress on Rock Mechanics, 3, pp. 7-13, 1970.
5. Engineering News Record, "Upside Down Retaining Wall Makes Project Feasible," Vol. 196, No. 16, p. 43, April 15, 1976.
6. Gould, J. P., "Lateral Pressures on Rigid Permanent Structures," State of the Art Papers, ASCE Specialty Conference on Lateral Stresses and Earth Retaining Structures, pp. 219-267, 1970.
7. Henkel, D. J., "Geotechnical Considerations of Lateral Stresses," State of the Art Papers, ASCE Specialty Conference on Lateral Stresses and Earth Retaining Structures, pp 1-49, 1970.
8. Littlejohn, G. S. and Bruce, D. A., "Rock Anchors Design and Quality Control," Proceedings. Sixteenth Symposium on Rock Mechanics, University of Minnesota, Minneapolis, Sept., 1975.
9. Merriam, R., "Portuguese Bend Landslide, Palos Verdes Hills, California," Journal of Geology, Vol. 68, p. 140, 1960.
10. Morgenstern, N. R. and Eisenstein, Z., "Methods of Estimating Lateral Loads and Deformations," State of the Art Papers, ASCE Specialty Conference on Lateral Stresses and Earth Retaining Structures, pp 51- 102, 1970.
11. Pilot, G., "Study of Five Embankment Failures on Soft Soils," Proceedings, ASCE Specialty Conference on Performance of Earth and Earth-Supported Structures, Vol. 1, pp 81-100, 1972.
12. Reti, G. A., "Slope Stabilization by Anchored Retaining Wall," Civil Engineering, p. 49, April, 1964.
13. Seed, H. B. and Whitman, R. V., "Design of Earth Retaining Structures for Dynamic Loads," State of the Art Papers, ASCE Specialty Conference on Lateral Stresses and Earth Retaining Structures, pp 103-147, 1970.
14. Shannon & Wilson, Inc., "Report on Foundation Studies PSHI, Seattle Freeway, Olive Way to East Galer Street," to Director of Highways, Washington State Highway Commission, 1963.
15. Terzaghi, K., "Large Retaining Wall Tests. I. Pressure of Dry Sand," Engineering News Record, Vol. 112, pp 136-140, 1934.

SLIDE STABILIZATION 4TH ROCKY FILL, CLINCHFIELD RR

by

G. L. Tysinger [1]

INTRODUCTION

A tieback wall stabilized an active slide which threatened the busy 25 million gross-ton mile main line of the Clinchfield Railroad through the Blue Ridge Mountains in North Carolina. The tiebacks provided full lateral support to the sliding mass. This external force was distributed to the soil by H-piles coated with coal tar epoxy. Treated oak lagging supported the cut face exposed between the H-pile soldier beams. The external restraint force increased the factor of safety against sliding from about 1.1 to 1.4.

The Clinchfield RR selected the tieback wall because tiebacks were ideally suited to the requirements of the project. These requirements were as follows:

1. Maintain uninterrupted rail traffic during construction.

2. Prevent slides that may result from construction operations.

3. Support surcharge from 15,600 ton (138,785 kN) unit trains.

4. Maintain track stability independent of the stability of the slope below the restraint system.

5. Regrade the embankment adjacent to the tracks to a 2:1 slope.

6. Provide embankment drainage between the sidehill cut face and the restraint system.

7. Limit construction to the dry season (Sep. through Dec.)

A geotechnical investigation provided shear surface location, shear strength parameters, water table elevations, stability analysis and slide control recommendations. Post construction monitoring shows that the slide has been stabilized.

---

[1]Construction Engineer, Clinchfield RR., Erwin, TN

HISTORY

The section of the Clinchfield RR. between Altapass and Ashford, NC was built between 1906 and 1908. It descends approximately 900 ft (270 m) across the slide prone eastern slope of the Blue Ridge Mountains. A 1.2 percent compensated grade and 8 degree maximum curvature control this serpentine route for 20 miles (32 km). Sidehill cuts and fills, ravine and hollow crossings, tunnels and switchbacks form a tortuous route that requires 18 miles (29 km) of track to cover an airline distance of 2.3 miles (3.7 km). Uncontrolled fills that contained trees, tree stumps and boulders formed the numerous crossings of the steeply sloping ravines and hollows. Construction crews dumped fill on the hillside without subgrade preparation, fill compaction or key excavation. Seepage pressure has developed because prior drainage paths were obstructed or have been restricted with time and events.

The Fourth Rocky Fill at Mile Post 197.8, shown by Fig. 1, is a typical crossing. This fill was placed between two spurs, one crossed by a tunnel and one crossed by an open cut. The tunnel side contains shot rock from the tunnel to the power pole, Fig. 1. From that point it contains micaceous silty sand for some 180 ft (55 m). The embankment forms a track roadbed that joins the rock slope;

Fig. 1. Embankment at the Fourth Rocky Fill prior to construction of the tieback slide control wall.

hence, the fill blocked the natural drainage. The height of fill at trackside is approximately 40 ft. These conditions caused downhill movement of the embankment. This slide, active for many years, was fed by a trackside fill placed to fill the scarp continously formed by the movement. In 1974 the Clinchfield constructed a trackside rail pile wall that was tied across the tracks to other rail piles and to rock anchors. Rail piles were satisfactory so long as the slope was stable; hence, a continuous supply of fill was need to fill the scarp at the rail pile wall, Fig. 2. Clinchfield maintenance

Fig. 2. Trackside scarp along the rail pile wall at the Fourth Rocky Fill.

forces placed trackside fill that aggrevated the slide. The stage was set for a slope failure that would destroy the mainline track.

SITE INVESTIGATION

The site investigation provided the information necessary to locate the shear surface, to locate the pheratic surface and to determine the shear strength parameters. The goetechnical consultant directed the making of the borings and the installation of piezometers and slope indicators. Figure 3. shows the location of this work. Soil and rock samples were taken by core barrels, split spoon samplers and Shelby tubes. Eight of the borings were core drilled to determine bedrock weathering and structure; the other two borings were stopped when the drive sampler refused. Direct shear, triaxial drained and mechanical analysis tests were made on Shelby tube samples taken from the shear zone.

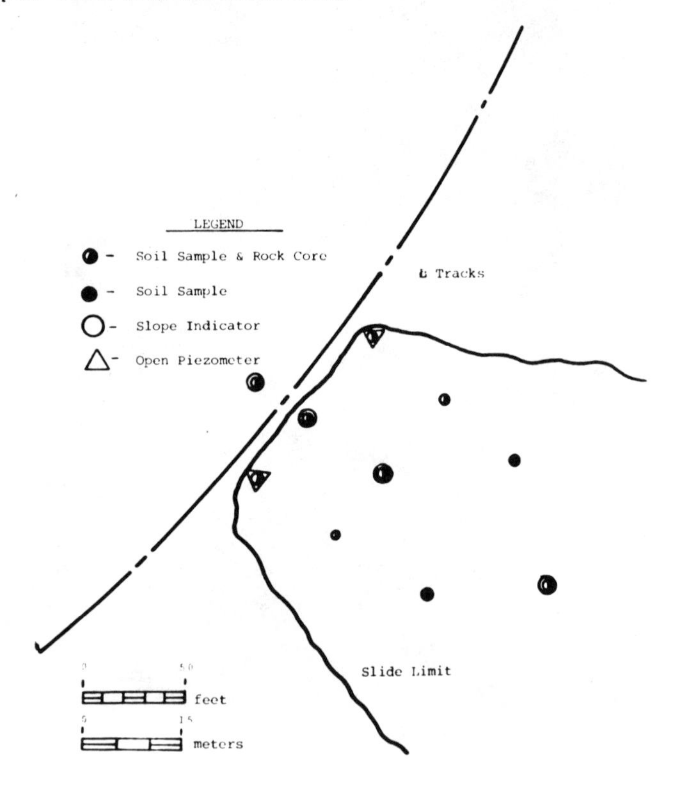

Fig. 3. The layout for the subsurface investigation at the Fourth Rocky Fill, Clinchfield, RR.

The typical slide section, Fig. 4, was 40 ft (12 m) of silty, micaceous sand overburden on weathered mica schist. The lower 4 to 10 ft (1 to 3 m) was residual soil and the remainder was fill. The shear surface followed closely the contact between the fill and residual soil and dipped at 30 degrees. The elevation of the pheratic surface varied, during the wet season it was approximately 10 ft (3 m) above the shear surface and during the dry season it was within the bedrock. One flank of the slide formed against the boulders deposited from the tunnel waste. The other flank developed along a spur. A steep scarp, Fig. 2, developed at the rail pile wall. The zone of accumulation began some 250 ft (76 m) downslope from the scarp and contained numerous springs during the wet season.

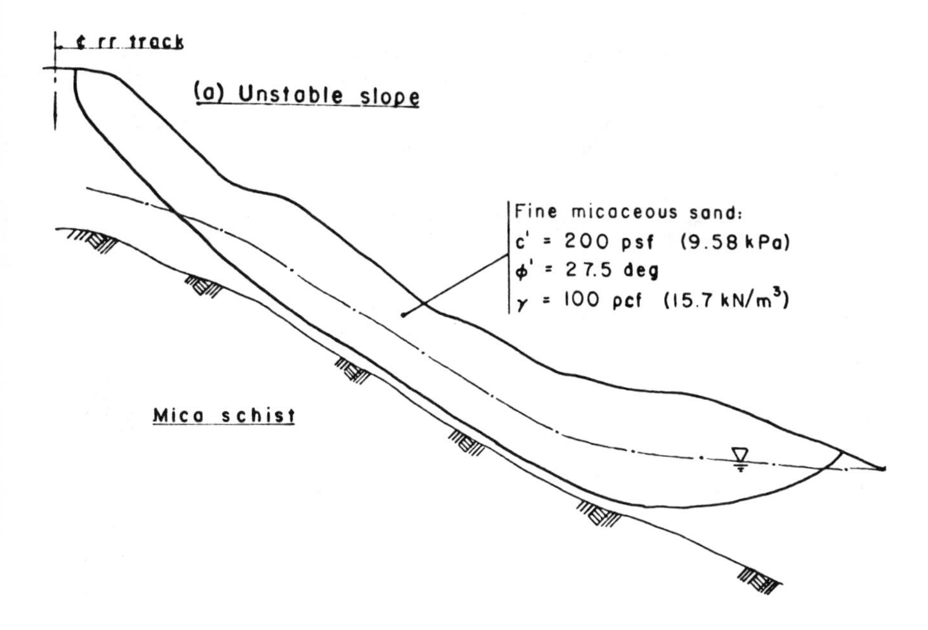

Fig. 4. Typical slide section the Fourth Rocky Fill, Clinchfield RR.

The soil along the shear surface behaved as a loose sand with slight cohesion; 0 to 15 percent passed the US No. 200 sieve. Laboratory tests measured effective shear strength parameters of 200 psf (9.68 kPa) for cohesion and 27.5 degrees for the angle of internal friction. The weathered surface of the mica schist was generally parallel to the 30 degree inclination of the shear surface. Foliation in the schist dipped at about 30 degrees into the slope, and a set of high angle joints normal to the foliation plane influenced the formation of the weathered surface. Since the pheratic surface dropped into the bedrock during the dry season, seepage pressure existed on the shear surface only during the late winter, spring and early summer.

EVALUATION

The Fourth Rocky Fill moved continously; however, the rate of movement increased during the wet season. Stability analysis using the methods of Bishop (1) and Janbu (2) gave a factor of safety against failure of 1.0 to 1.1 based on the following:

1. Shear surface shown by Fig. 4.

2. Cohesion equal to 200 psf (9.68 kPa); angle of internal friction equal to 27.5 degrees (effective shear stress parameters).

3. Pheratic 8 ft (2.5 m) above the shear surface.

For normal weather, the stability of the slope was marginal, and an intense storm during the wet season could cause failure. Much of the shear surface, Fig. 4, developed on a plane surface; hence, the restraint force required for a given factor of safety against sliding would decrease as the point of application is moved closer to the tracks.

The Fourth Rocky Fill site was analysed as an infinite slope, Fig. 5, by Janbu's (2) method. When fully saturated, this slope would become unstable at overburden depths greater than 10 ft (3 m). This condition lead to two conclusions. First, continued dumping of trackside fill aggrevated the already unstable condition. Second, the soil downslope from the external restraint system was susceptible to slides; hence, the restraint force must be distributed from the ground surface to the bedrock.

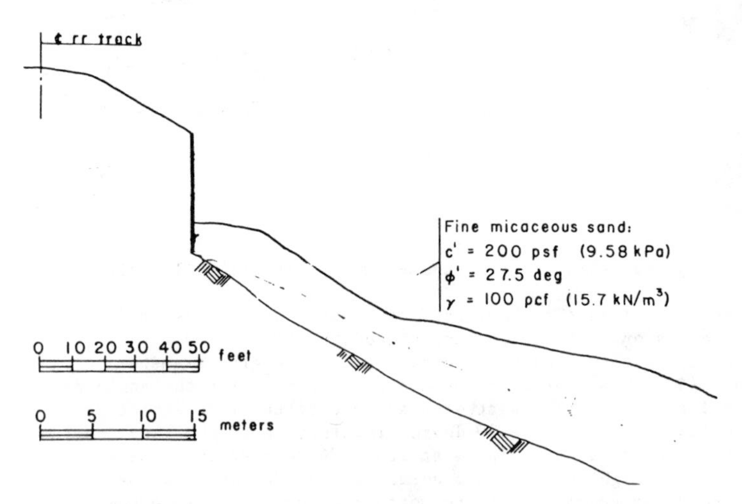

Fig. 5. Fourth Rocky Fill slope analysed as an infinite colluvial slope.

This active slide was well suited for tiebacks. The existing mainline tracks made regrading impossible. Free standing walls would require deep, shored exvatations that threatened stability and cost as much as a tieback wall. In this region horizontal drains clog and became unreliable. A tieback wall constructed downslope from the tracks, Fig. 6, satisfied the project requirements and eliminated interference between tiebacks and the rail pile wall. The tiebacks applied the full restraint force; the wall extended to the bedrock because intense storms during the wet season would cause slope failure below the wall.

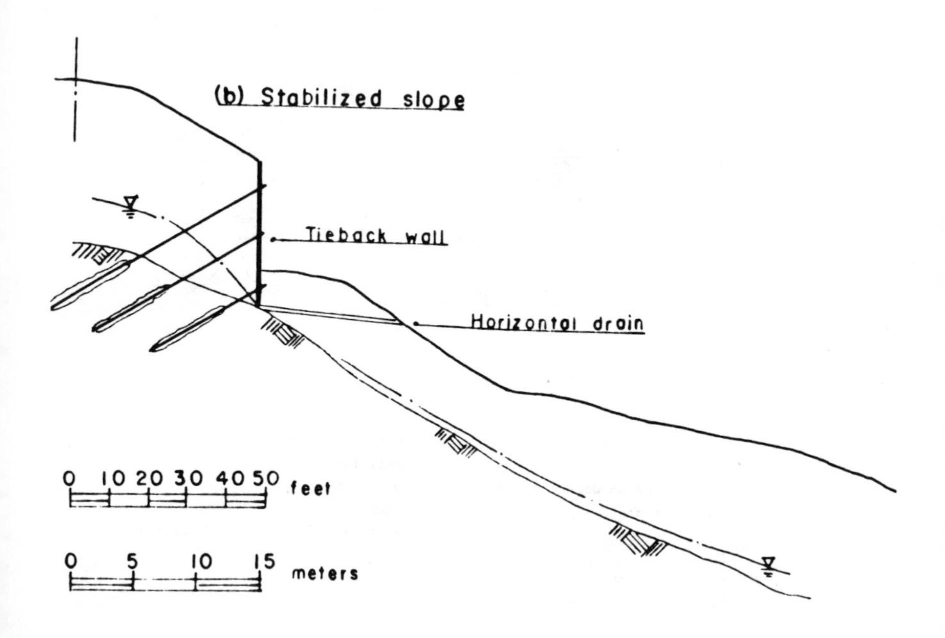

Fig. 6. Design concept for the tieback slide control wall at the Fourth Rocky Fill, Clinchfield RR.

DESIGN

The design involved permanency of the restraint system and calculation of the tieback restraint force. The tieback wall was designed as a permanent structure, and all tiebacks were tested to insure that they would maintain the restraint force. The restraint force required to stabilize the slope, Fig. 6., at the design factor of safety against sliding was calculated by Janbu's method (1).

Permanency. Permanent tiebacks were 1-3/8 in (35 mm) Dywidag bars fully coated with heat applied epoxy. The unbonded length of the tendon was covered by heat shrink plastic sheath that in turn was covered by polyethlyene bond breaker. Cement grout anchored the tieback in the mica schist bedrock. The anchor length was centralized in the drill hole to insure that the tendon was covered by at least 0.5 in (12mm) of portlant cement grout. A PVC pipe filled with grease and attached to the anchor head bearing plate protects the anchor head end of the tendon from corrosion, Fig. 7. Tape-coat moldable sealent

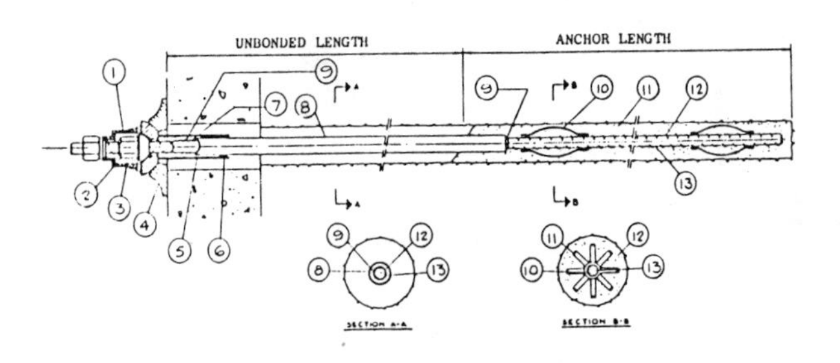

LEGEND:

1. Anchorage Cover
2. Nut
3. Anti-corrosion Grease
4. Bearing Plate
5. Trumpet
6. Seal
7. Anti-corrosion Grease or Grout
8. PVC or Polyethylene Tube
9. Heat Shrinkable Tube
10. Centralizer
11. Anchor Grout
12. Tendon
13. Electrostatically Applied Epoxy

Fig. 7. Tieback corrosion protection details

covered the exposed part of the anchor head. The H-pile soldier beams and wales were coated with coal tar epoxy. Treated wood lagging covered the cut face between soldier beams. Wood lagging requires regular maintenance; hence it was placed on the exterior flange of the soldier beam to simplify replacement. Enka-drain, a preformed drain path material, was installed behind the lagging for chimney drains. These drains were connected to subdrains that daylighted down slope from the wall. This feature will keep the water table below the shear surface. Figure 8 shows these wall details.

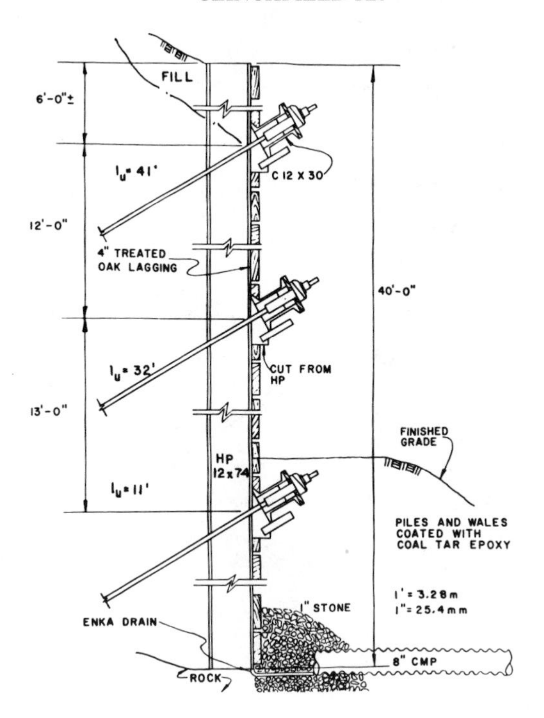

Fig. 8. Tieback slide control wall details, Fourth Rocky Fill, Clinchfied RR.

Performance and proof tests were performed on all tiebacks to insure that the tiebacks would permanently carry the design restraint load. Creep tests are not necessary for tiebacks anchored in rock.

Restraint Force. Limit equilibrium analysis predicts reasonable values for the factor of safety against sliding and restraint force because the soil along the shear surface is cohesionless (2) and (3). Janbu's method (1) was used because it is suitable for a programmable calculator and allows iterative solution for the horizontal restraint force required for equilibrium at a given factor of safety. The basic equation for Janbu's Generalized Procedure of Slices is

$$F = \frac{\sum_a^b \Delta x(c' + (p + t - u)\tan\phi')\frac{(1 + \tan^2 \alpha)}{(1 + \tan\alpha\tan\phi'/F)}}{E_a - E_b + ((p + t)\Delta x\tan - \Delta Q)} \quad \text{Eq. 1}$$

A detailed discussion of Eq. 1 and its solution are presented in (1) and are not repeated here. Fig. 9 shows the restraint system for this

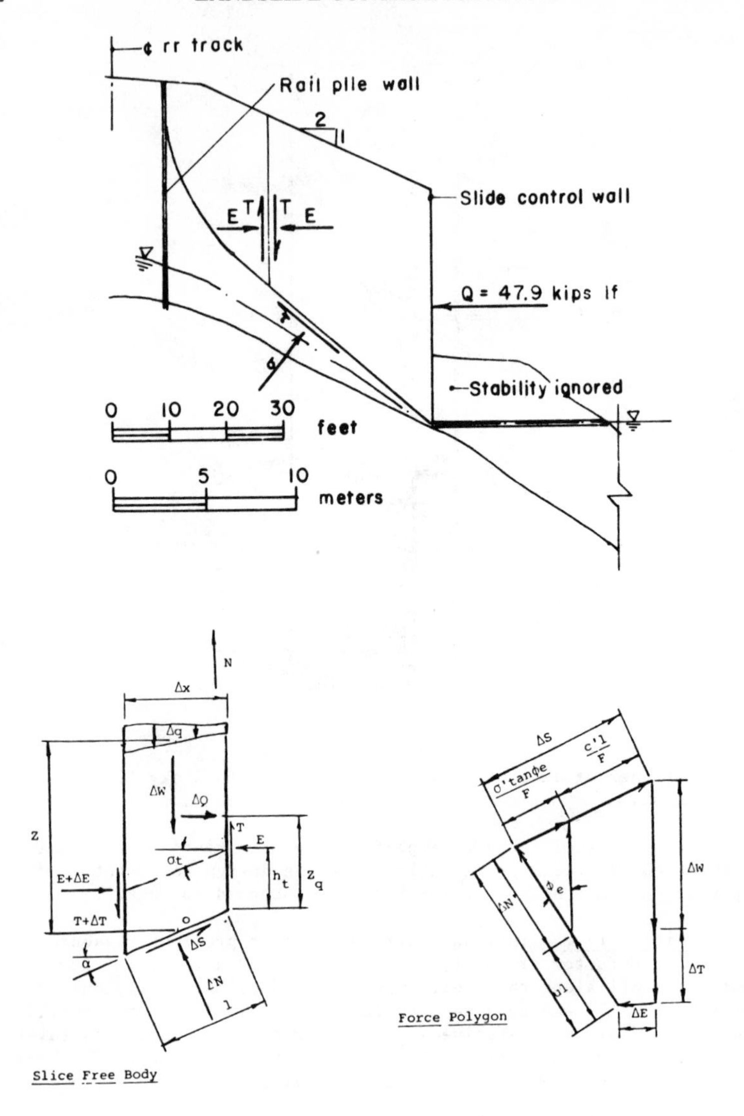

Fig. 9. Model for stability analysis, Fourth Rocky Fill, Clinchfield RR.

tieback slide control wall. The design factor of safety of 1.4 required a horizontal horizontal restraint force of 48 kips per LF (65 kN per m). HP 12 x 74 soldier beams, 7 ft (2 m) on centers and driven to refusal on the bedrock, distributed the horizontal restraint force as shown by Fig. 10. The treated wood lagging carries only that portion of the restraint force that cannot be carried in bearing

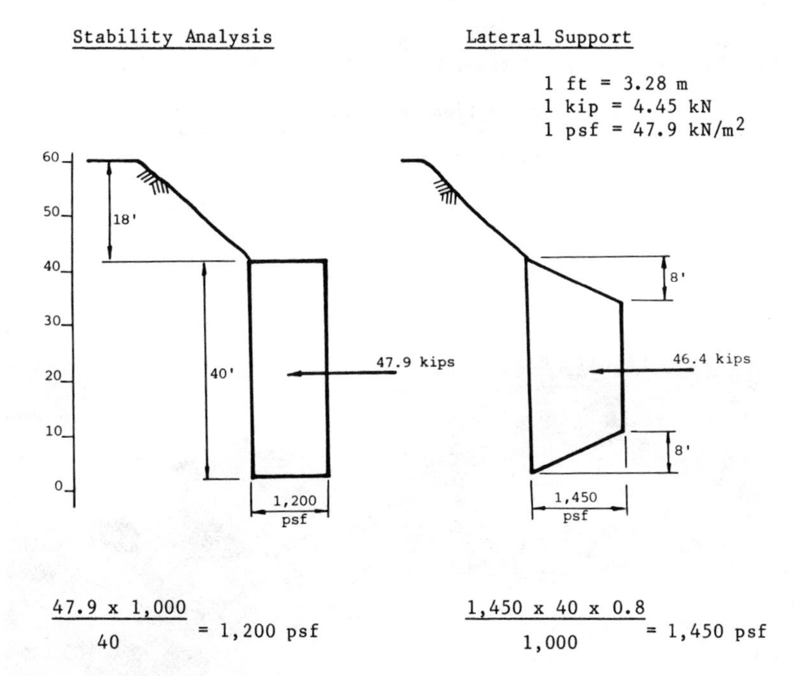

The critical case, stability analysis, determined the design restraint force. These apparent pressure envelopes place the restraint force at the mid height of the cut; this minimizes ground movements.

Fig. 10. Design earth pressure envelopes for Fourth Rocky Fill tieback slide control wall.

between the soil and the soldier beam. The lagging gives local support to the soil between the soldier beams; hence it is a secondary part of the restraint system. The tiebacks were installed at 30 degree angle, Fig. 8, and the resulting vertical tieback load was carried by the soldier beams in bearing on the bedrock. Soldier beams and lagging went to bedrock for support of the vertical tieback load and for erosion back of the wall in the event of a downslope slide. If not for these requirements, the wall could have been terminated about 2 ft (1 m) below the bottom tieback tier.

CONSTRUCTION

The experience gained from construction of this tieback wall is summarized below.

1. Slope movement continued during pile driving, but it stopped as soon as the first tier of tiebacks were tensioned.

2. The soldier beams moved up to 8 in (200 mm) into the soil because the top tier of tiebacks were tensioned against fill.

3. Ten percent of the tiebacks failed because grout was lost in shot rock placed on bedrock. All tiebacks should be tested, and the contractor should be made responsible for tieback performance.

4. The completed slide control wall, Fig. 11, shows that H-piles can be driven to the tolerences required for a satisfactory wall.

Fig. 11. Completed slide control wall, Fourth Rocky Fill, Clinchfield RR.

BEHAVIOR

The post construction performance of the tieback wall has been monitored by load cells and by reference tapes attached to the top of the soldier beams. Six load cells were installed, two on each tier in two vertical lines in the center portion of the wall. The reference tapes measured horizontal movement of 24 of the 27 soldier beams.

Tieback performance has been monitored by load cells for 12 months. Table 1 gives the results of these measurements; changes

TABLE 1.
LOAD CELL READINGS

	Tieback Load (kips)			
Date	12-1	12-2	12-3	Total
Lockoff	93.7	102	100.2	295.9
1/26/81	102.8	130.2	101.5	334.5
Change	+ 3%	+ 3%	+ 6%	+ 4%
3/16/81	105.6	134.3	107.4	347.3
Change	- 5%	+ 2%	+ 4%	+ 0%
6/ 4/81	100.5	136.6	111.9	349.0
Change	- 3%	- 1%	+ 1%	+ 1%
7/15/81	97.7	135.8	112.6	346.1
Change	- 3%	- 1%	+ 3%	0%
12/14/81	95.1	134.5	115.6	345.2
Total change	- 8%	+ 3%	+14%	+ 5%
	16-1	16-2	16-3	Total
Lockoff	102.3	102	94.4	298.7
1/26/81	99.5	109.2	100.1	308.8
Change	+1%	+ 4%	+ 2%	+ 2%
3/16/81	100.5	113.1	102.4	316.0
Change	- 1%	+ 2%	+ 4%	+ 2%
6/ 4/81	99.2	115.6	106.0	320.8
Change	- 1%	0%	+ 1%	0%
7/15/81	97.7	115.5	107.0	320.6
Change	- 2%	0%	+ 1%	0%
12/14/81	95.7	115.7	108.3	319.7
Total change	- 3%	+ 6%	+ 8%	+ 4%

Load cell 12-1 is located at soldier beam 12 top tier, 12-2 is located at soldier beam 12 middle tier, etc. 1 kip = 4.45 kN

are the quotent of the difference between successive load cell readings divided by the 1/26/81 load times 100. Tieback loads in the top tier decreased because the soldier beams developed bearing capacity of the top 6 ft (2 m) of compacted fill, Fig. 8, by inward movement. This is a common occurrance for tiebacks tensioned against backfill. Loads in the middle and bottom tier increased because the soil behind the wall strained to develop shear strength that was required for stability of the cut face. Bottom tier tieback loads increased more than middle tier loads because the middle tier unbonded length was about 3 times that of the bottom tier, Fig. 8. Load cells 12-2 and 16-2 were not read at the time of lock off; these lockoff loads were computed from the jack pressure. Soil movement that occurred during the excavation from the middle tier to bed rock probably caused the increase. Depth of bedrock below the bottom tier was 10 ft (3 m) and 6 ft (2 m) for soldier bearm 12 and 13 respectively.

The pile movement tapes were referenced after all tiebacks were installed and tensioned. Movements were as follows:

Lock-off to placing of berm (1/22/81)	0.018 ft (5.5 mm)
1/22/81 to 7/1/81	0.003 ft (0.9 mm)

CONCUSIONS

1. Tiebacks will stabilize an active slide.

2. The reaction element distributes the tieback force to the cut face. This element behaves like a shallow footing; hence, lagging receives load when the soil or rock under the reaction element fails in bearing.

3. The lockoff load in the tieback lowers the level to which the shear strength must be mobilized to stabilize the slope.

4. The restraint force can be estimated by limit equilibrium analysis with effective shear strength parameters.

5. Tieback slide control walls will apply the design restraint force without penetration of the shear surface.

6. Tiebacks will lose lockoff load when tensioned against compacted backfill.

7. The unbonded length controls the load increase from soil movement.

Appendix I - References

1. Janbu, N., Slope Stability Computations, Embankment Dam Engineering (Hirchfeld, R. C. and Poulos, S. S., eds), Wiley, New York, 1973, pp. 47-86.

2. Morgenstern, N. R. and Sangrey, D. A., Methods of Stability Analysis, Landslides Analysis and Control (Schuster, R. L. and Krizek, R. J., eds), Special Report 176, National Academy of Sciences, Washington, D. C., 1978, pp. 155-169.

3. Tavenas, F., Trak, B., and Leroueil, S., Remarks on the Validity of Stability Analysis, Canadian Geotechnical Journal, Vol. 17, No. 1, 1980, pp. 66-73.

Appendix II - Notation

The following symbols are used in this paper:

- c = cohesion, effective stress
- E = horizontal force on slice interface
- F = factor of safety
- l_u = unbonded length
- q = surcharge pressure
- Δq = surcharge pressure over slice
- p = vertical pressure on the shear surface
- Q = tieback force, horizontal component
- ΔQ = change in tieback force across the slice
- T = vertical shear force on slice interface
- ΔT = change in vertical shear force across the slice
- t = $T/\Delta x$
- u = excess pore pressure
- W = weight of the sliding mass
- ΔW = weight of a slice
- Δx = width of a slice
- α = inclination of the shear surface with respect to the horizontal
- ϕ' = angle of internal friction, effective stress
- σ' = effective normal stress on the shear surface
- τ' = effective shear stress on the shear surface

Subscripts:

- a - moving mass boundary at the scarp
- b - moving mass boundary at the toe
- e - value at limit equilibrium

TANGENT PILE WALL, EDMONTON CONVENTION CENTRE

Lawrence A. Balanko, P.Eng.[1], Norbert R. Morgenstern, P.Eng.[2], and Rudy Yacyshyn, P.Eng.[3]

INTRODUCTION

The City of Edmonton has actively planned development of a Convention Centre since about 1975, when initial geotechnical evaluation of a favoured site was undertaken. The site is located near the City centre and is bounded by a major thoroughfare, Jasper Avenue on the north, and the North Saskatchewan River valley slopes to the south. A major historic landslide exists immediately adjacent to the east side of the site on Grierson Hill Road, which traverses through the lower portion of the site in its ascent of the valley slope.

In spite of the aesthetically pleasing and favourable location for such a facility, it was recognized that the site characteristics, particularly its proximity to an active slide, would necessitate a detailed geotechnical review and analysis of development feasibility. Accordingly, EBA Engineering Consultants Ltd., in cooperation with Dr. N.R. Morgenstern, were commissioned by the City of Edmonton to undertake a geotechnical assessment of site stability, and the provision of design requirements for the proposed development. This work commenced in August 1978, with the start of construction, in March 1980, of a major permanently anchored earth-retaining wall supporting vertical excavation slopes on three sides of the building.

This paper summarizes the geotechnical evaluations undertaken and demonstrates their influence on the design ultimately selected for construction. It also reviews the performance of the wall through construction, and over the past year, on the basis of the instrumentation installed for that purpose.

Note:

[1]Principal, EBA Engineering Consultants Ltd., Edmonton, Alberta, Canada

[2]Professor of Civil Engineering, Department of Civil Engineering, University of Alberta, Edmonton, Alberta, Canada

[3]Project Director, City of Edmonton, Real Estate and Housing Department, Edmonton, Alberta, Canada

SITE ASSESSMENT

GEOLOGY - Regional geology plays an important role in the stability of the Edmonton Convention Centre site. Kathol and McPherson (1975) describe the geology as a succession of glacio-lacustrine sediments, glacial till and Saskatchewan gravels and sands overlying soft bedrock.

The glacio-lacustrine sediments comprise slightly over-consolidated silt and clay; whereas the glacial till is a highly over-consolidated, well-graded, mixture of clay, silt, sand and gravel sizes. The gravels and sands are dense and were deposited by Tertiary period streams flowing on the preglacial bedrock landforms.

Most significant of the bedrock strata at the site is the Edmonton Formation, deposited during the upper Cretaceous Period. This formation consists of fine-grained bentonitic sandstone and siltstone, interbedded with silty claystone. Coal and thin bentonite seams are common throughout the formation. Significant features of the upper bedrock units are thin zones of sheared or weakened materials. These have resulted from unloading and straining of the bedrock layers due to valley rebound, as discussed by Matheson and Thomson (1973), and from glacial action.

These weak zones and layers of bentonite have contributed to major landslides in the Edmonton area, as occurred east of the site in the late 1800's. Although the slide likely was activated by river erosion at the toe, early coal mining activity was demonstrated by Hardy (1961) to have altered the groundwater regime and contributed to continued instability. Accordingly, a drainage gallery was installed along Grierson Hill Road in the 1950's to stabilize the slide area. Although this has had a stabilizing effect on the adjacent slide, some movements have persisted. The site, however, has not experienced recent movements other than minor surficial sloughing. The site location, historical slide activity and features of significance to site stability are depicted on Fig. 1.

EXPLORATION PROGRAM - Although substantial geotechnical data existed for the area, it was decided to undertake detailed subsurface exploration of bedrock strata beneath the site. This decision was based on the perception that site stability was related to the presence and nature of weak zones in the bedrock. Therefore, deep continuous core sampling of the bedrock was undertaken using triple-tube core barrels. Additional auger holes were drilled to define stratigraphic detail in the overburden. Borehole locations are shown on Fig. 1.

All core was carefully examined and logged in the field and laboratory. Bentonite layers and pervasively sheared zones, displaying shiny slickensides, were identified. Their locations were compiled on stratigraphic sections, together with potential weakness and failure zones from the adjacent slide area, as shown on Fig. 2.

Instrumentation consisting of pneumatic piezometers and tiltmeters was also installed as part of the field program. Observed groundwater levels are shown on Fig. 2. The tiltmeters revealed no movements and, hence, indicated stability in the upper and lower site areas.

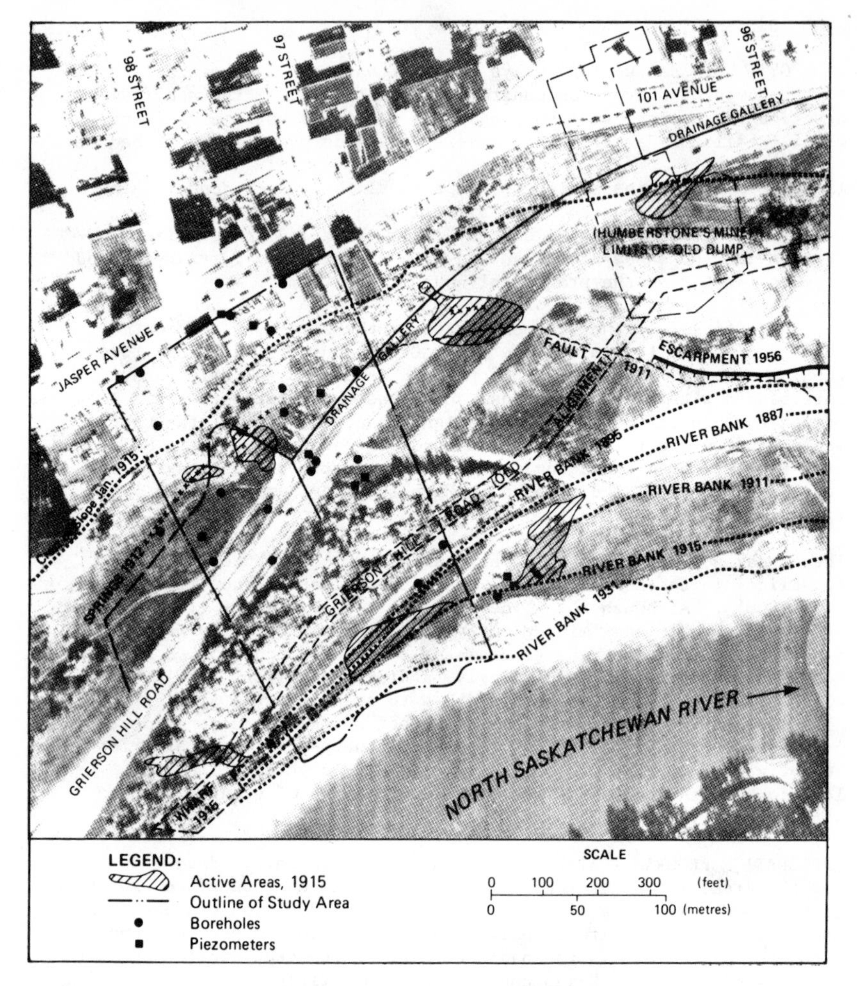

FIG. 1 - SITE DATA - GRIERSON HILL - EDMONTON CONVENTION CENTRE

LABORATORY TESTS - The major emphasis of the laboratory program was a detailed core examination, and identification and mapping of zones of weakness. Samples from these zones were subjected to slow, drained, direct-shear tests. A range of effective strength parameters was obtained from these tests as shown on Fig. 3.

The strength test results were reviewed relative to other work undertaken on these materials by Thomson (1970, 1971), Eigenbrod and Morgenstern (1972), Locker (1973), and Thomson and Yacyshyn (1977). Based on these sources and the test results, the parameters presented in Table 1 were selected for stability analyses of the site.

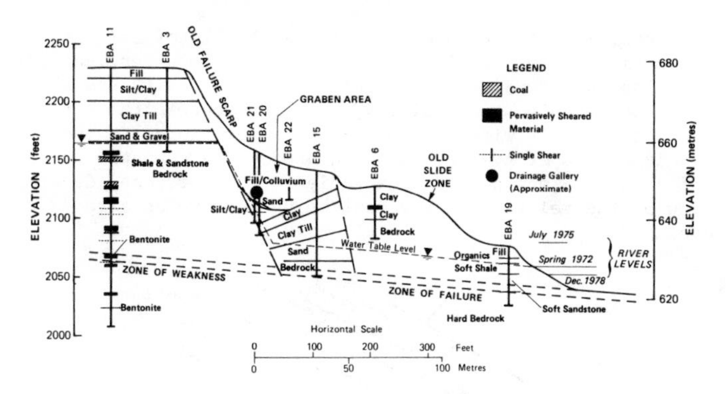

FIG. 2 - SITE STRATIGRAPHY, NORTH-SOUTH SECTION

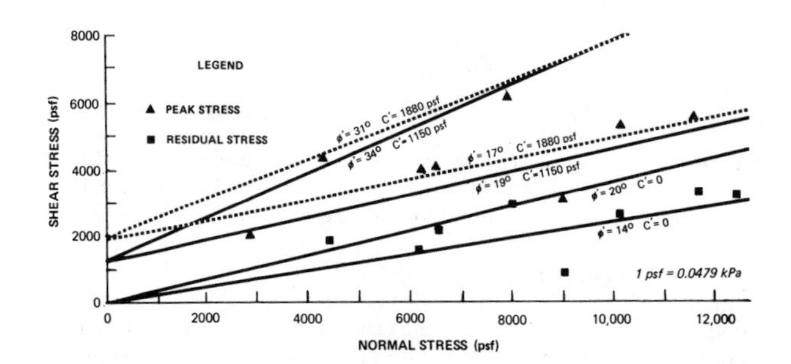

FIG. 3 - EFFECTIVE STRENGTH PARAMETERS - SHEAR ZONE MATERIALS

TABLE 1 - STABILITY ANALYSES - STRENGTH PARAMETERS

MATERIAL	ANGLE OF SHEARING RESISTANCE (degrees)	COHESION (psf)	COMMENTS
Fill, Silt, Clay	24	0	
Clay Till	40	0	
Sand and Gravel	40	0	
Shale and Scarp	25	1000	Cross bedding
In place Bentonite	14	0	Zone of weakness
Previously sheared Bentonite	8	0	Zone of failure

1 psf = 0.0479 kPa

STABILITY ANALYSES - Numerous limit equilibrium stability analyses were performed for typical slope sections, and groundwater conditions, as shown on Fig. 2. The analyses, using the Morgenstern-Price computer program, assessed both natural stability and that of construction-modified slopes. Results of the parametric analyses performed are shown on Fig. 4. Potential changes in groundwater levels were also modelled to assess sensitivity of stability to piezometric conditions.

From the analyses, it was concluded that the stability of the natural slope was insufficient for development of the site. However, unloading of the upper slope was considered feasible to increase site stability and render it developable. Accordingly, a design wherein the structure would be founded entirely within a cut in the upper slope, to the northerly limit of Jasper Avenue, was advocated. The computed factor of safety of about 1.4 was deemed acceptable for the proposed development, in view of the relative improvement being made to an already stable slope.

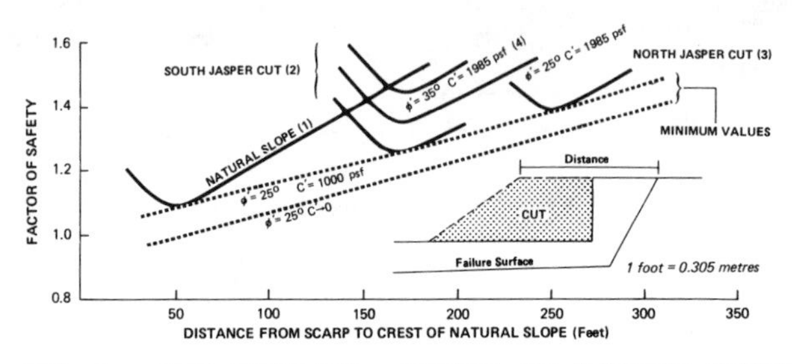

FIG. 4 - SITE STABILITY - NATURAL AND MODIFIED SLOPES

Development on lower slope sections was not recommended as this area was concluded to be an old land slide, possessing only marginal stability. Improved stability could be achieved only with difficulty.

CONCEPTUAL DESIGNS - Structural space requirements dictated an excavation depth of about 65 ft (19.8 m). Moreover, the vertical excavation walls would need permanent support. Regional studies had shown that long-term softening was an important mechanism leading to slope instability. Therefore, the support system was needed not only to support adjacent ground, but also to impress upon the adjacent ground substantial pressures in order to inhibit long-term softening. It was concluded that the most suitable system was a tangent pile wall supported by sequentially installed permanent soil anchors.

Permanent anchors were believed feasible in three strata on site. However, lack of local experience with permanent anchors in "weak soils and rock", necessitated some conceptual anchor design assumptions. Anchor bond capacities were, therefore, conservatively estimated for the glacial till, Saskatchewan gravels and sands, and Edmonton Formation bedrock, as 1000 psf (48 kPa), 3550 psf (170 kPa), and 2820 psf (135 kPa), respectively. Anchors were assumed to develop capacity in direct proportion to bond area.

Further conservative assumptions were made regarding soil pressure on the wall. In this respect, a rectangular stress distribution, of 0.4 γH, was assumed to represent the applied soil pressure. Piezometric pressures were not considered in the design due to the free draining nature of a tangent pile wall. Moreover, the groundwater conditions observed were clearly of a perched nature.

Based on the assumptions, seven rows of anchors, at 9.0 ft (2.7 m) vertical and 7.0 ft (2.1 m) horizontal spacing, were required.

Resulting anchor loads, up to 250 kips (1112 kN), would necessitate either very long or large diameter anchors. Nevertheless, the conceptual designs were determined to be within the realm of feasibility. Accordingly, a full-scale anchor test program was recommended to permit confirmation of feasibility and refinement of design parameters.

ANCHOR TEST PROGRAM

Only two strata were tested; the glacial till and the Edmonton Formation bedrock. The Saskatchewan gravels and sands were considered to be of too limited depth to test. Therefore, test locations were selected on the upper and mid-slope portions of the site, from which anchors were installed, at 30 to 35 degrees from the horizontal, into the glacial till and bedrock, respectively. Eight and nine anchors were installed in the upper and mid-slope sites, respectively.

TEST ANCHOR TYPES - Four 24 in (610 mm) diameter anchors were installed in the glacial till. These anchors were fabricated of six high-strength 1.25 in (32 mm) diameter thread-bars. The purpose of testing such a large anchor was to assess the effects of area on bond capacity. The remaining four test anchors in the till were 8 in (203 mm) diameter, multi-strand, assemblies.

A combination of 8 in (203 mm) and 6 in (152 mm) diameter multi-strand anchors were installed at the bedrock test site. These were fabricated of 0.62 in (16 mm) stabilized steel strands, generally as shown in Fig. 5. Bond zone lengths of 20 ft (6.1 m), 30 ft (9.1 m) and 40 ft (12.2 m) were selected for testing.

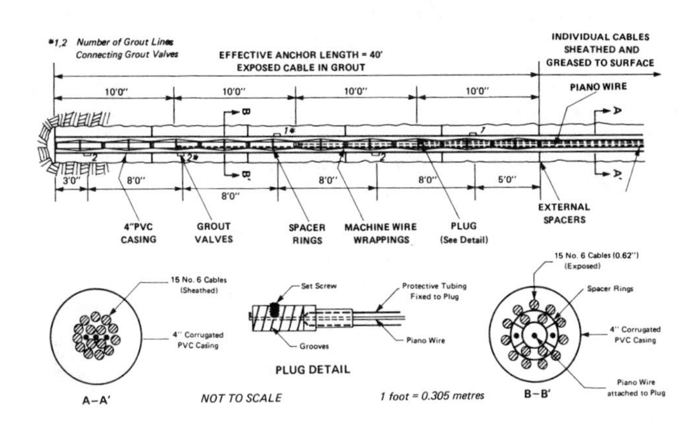

FIG. 5 - TYPICAL MULTI-STRAND TEST ANCHOR

TEST ANCHOR INSTALLATION - All small diameter anchors were drilled with an AP1000 CSR rotary drill rig, employing double wall casing and reverse circulation of air and water mist for return of cuttings. The large anchors were drilled with a conventional piling rig.

Primary grouting consisted of pumping a cement grout, in the proportion of one bag of cement [87 lb (40 kgm)] and 4.5 gal (20.5 L) of water, into the anchor hole. Once it was determined that grout loss was not occurring, the free-stress length of the hole was flushed of cement grout with a bentonite/water slurry.

Secondary pressure grouting was performed in multiple stages. The intent was to achieve approximately 9 cu ft (0.25 m^3) of grout injection per valve. Typical valve arrangements are shown in Fig. 5. Secondary grouting success was variable and most anchors were grouted in three or more stages.

Grouting pressures ranging from 300 to 1300 psi (2069 to 8963 kPa) were required to break the grout bond and initiate grout take. The purpose of the pressure grouting was to increase the bond capacity of the soil/anchor system as noted in the work of Ostermayer (1974) and Littlejohn and Bruce (1977).

ANCHOR LOAD TESTING - Most of the anchors were loaded with centre-pull hydraulic jacks. Load was measured with an accurately calibrated 800 kip (3560 kN) load cell. Extensometer wires, affixed to the tendons and within the anchor bond zone, were routed over a measuring frame to monitor displacement and debonding effects. However, difficulties with binding of bond zone extensometers rendered these latter measurements of little value.

Incremental loading stages were selected to evaluate total, elastic and permanent displacements, up to ultimate load capacity. Time-displacement measurements were made to assess anchor creep rate. Test procedures were similar to those prescribed in the German Code (Draft DIN4125:1974) for permanent anchors, as highlighted by Littlejohn and Bruce (1977).

Load-displacement curves were obtained for all anchors, as typified by Fig. 6, for an 8 in (203 mm) anchor in bedrock.

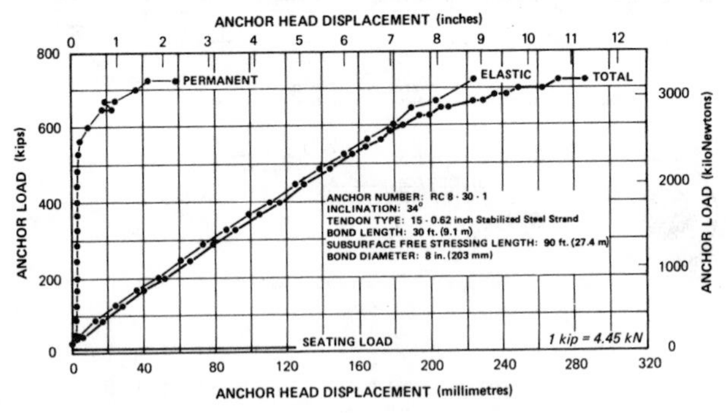

FIG. 6 - LOAD-DISPLACEMENT - ANCHOR RC8-30-1

The maximum load achieved was 720 kips (3204 kN). Fig. 7 shows creep-displacement curves for the same anchor. Based on a creep rate failure criterion of 0.08 in (2 mm) per log cycle of time, the failure load is just under 600 kips (2670 kN), as apparent from Fig. 7, and more readily from Fig. 8.

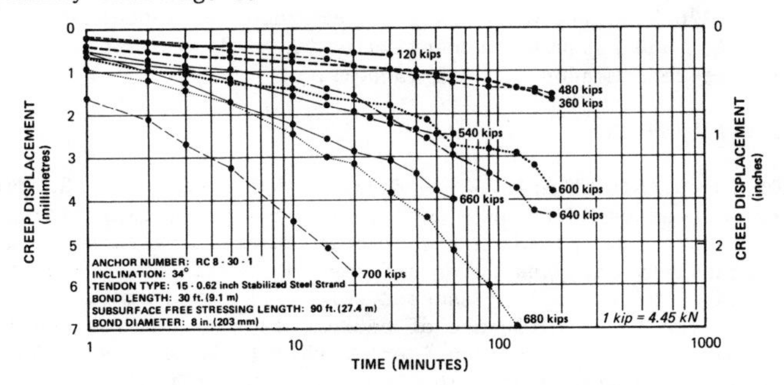

FIG. 7 - CREEP-DISPLACEMENT - ANCHOR RC8-30-1

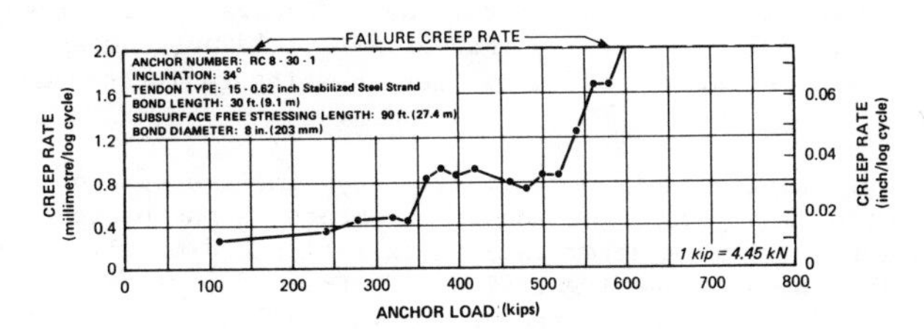

FIG. 8 - CREEP RATE - ANCHOR RC8-30-1

Several anchors were not tested to failure immediately. These were loaded to and locked-off at the estimated design load. Measurements of load were made with time to assess load relaxation effects. Since these anchors were exposed, temperature effects on anchor load were also monitored. Fig. 9 is a typical plot of load and temperature with time. The parallelism between load and temperature change is readily apparent. This behaviour was generally typical, with relaxation noted to be about 1% for bedrock anchors.

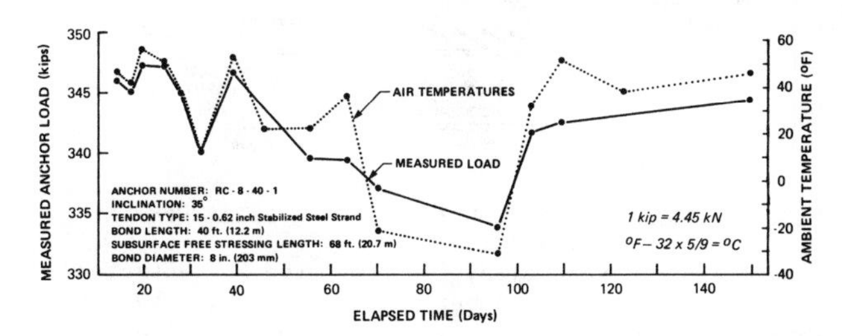

FIG. 9 - LOAD AND TEMPERATURE WITH TIME - ANCHOR RC8-40-1

ANCHOR TEST CONCLUSIONS - Based on observations of anchor fabrication and installation, and load test results, it was concluded that permanent double corrosion-protected, multi-strand, anchors were feasible for wall support. Accordingly, anchors of 12 in (305 mm) and 8 in (203 mm) nominal diameter, with design working loads of 250 kips (1113 kN) and 320 kips (1424 kN) were recommended for installation in the glacial till and bedrock, respectively. A bond zone length of 40 ft (12.2 m) was recommended for all anchors.

FINAL DESIGN ASSESSMENT

DEFORMATION AND STABILITY ANALYSES - Final design details of the wall and anchors were assessed through deformation analyses, employing a finite element program, based on the principles of linear elasticity. Additional stability analyses were performed on various wall/anchor configurations. The variables assessed were wall depth and anchor bond zone location. The influence of bedrock shear zones was modelled in the deformation analyses by considering them to be transversely isotropic elastic materials with low horizontal shear resistance. This low shear resistance was simulated by assigning shear moduli equal to 1/10 and 1/1000 of the corresponding Youngs Modulus [typically 20 000 psi (137.8 mPa)]. To inspect for yielding, comparisons were made between available and mobilized shearing resistance in the weak zones.

Excavation unloading effects were modelled by single-step and multiple-step deformation analyses. Horizontal movements in the bedrock below the excavation level were determined to be influenced mainly by the three principal shear zones identified. Control of movements above a shear zone would necessitate extension of the wall below that zone. However, it was concluded that a wall 100 ft (30.5 m) deep, terminating below the middle shear zone, would be adequate for the facility. The final wall/anchor configuration recommended is depicted in Fig. 10, which also shows the design anchor loads and the equivalent horizontal stress distribution for the wall.

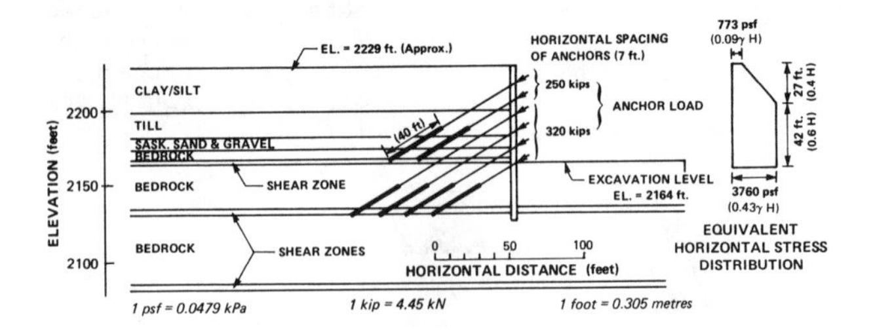

FIG. 10 - WALL/ANCHOR CONFIGURATION - FINAL DESIGN

Maximum horizontal shear zone slip, for simulated instantaneous unloading, was determined to be about 1.2 in (30 mm). The calculated maximum excavation heave was about 1.5 in (40 mm). Horizontal stress and shear stress with depth, was also obtained as output from the computer analyses and used as input to structural design of the wall.

Actual excavation unloading was simulated by a multiple-step analysis.

Similar output to the single-step analysis was obtained. However, the maximum horizontal displacement at depth was determined to be about 0.4 in (10 mm) and the excavation base heave about 1.2 in (30 mm). Because of progressive restraint provided by the anchors, these lesser displacements were deemed to be more representative of actual conditions. However, this was not found to be so, as discussed later.

CONSTRUCTION OF THE WALL AND ANCHORS

WALL DETAILS - The wall constructed consists of 312, 3.5 ft (1.07 m) diameter, concrete tangent piles installed to an average depth of about 103 ft (31.4 m) below upper-site surface elevation. Reinforcement varies with depth according to moment and shear capacity requirements. Total wall length is 1102 ft (335.9 m). The maximum depth of excavation adjacent to the wall is about 65 ft (19.8 m).

ANCHOR DETAILS - Final anchor design and spacing was essentially as proposed. However, an alternate was approved for the top two rows, and consisted of 120 kip (534 kN) capacity, single 1-3/8 in (35 mm) thread-bars spaced at 3.5 ft (1.07 m) horizontally. All other anchors were spaced at 7.0 ft. (2.1 m) horizontally. Six rows of anchors, spaced at 10.5 ft (3.2 m) vertically, were installed in the main excavation level. Two additional rows of anchors support the lower stepped-portion of the excavation, as shown in Fig 11.

FIG. 11 - EDMONTON CONVENTION CENTRE - EAST WALL

Of the 1061 anchors installed, 505 were single thread-bar and the balance were multi-strand anchors. Most multi-strand anchors were fabricated as shown in Fig. 12. Design capacity of this anchor type was 320 kips (1424 kN). Near the east and west wall extremities, a number of lower capacity anchors were required. These consisted of 5, 7, and 9 strands with capacities of 150 kips (668 kN), 200 kips (890 kN) and 250 kips (1113 kN), respectively.

All anchor holes were drilled with AP1000 CSR rotary drill rigs, as employed in the test program. The nominal hole size was 8 in (203 mm) diameter, at an average inclination of 30 degrees from the horizontal.

Grouting of anchors was undertaken in two stages. In the primary stage, the anchor bond zone was filled with a water/cement grout.

The water/cement ratio ranged from 0.4 to 0.44. Intraplast N (0.56% of the cement weight) was used as an additive. Although the interior of the thread-bar tendons was pre-grouted, all strand anchors were grouted, internally and externally, in place. This was done to prevent potential fracturing of the double corrosion protection during installation. The free-stress length of the anchors was filled with a bentonite/cement/water slurry. A mix proportion of 13.7 lbs (6.2 kgm) of cement, 5.7 lbs (2.6 kgm) of bentonite and 8 gal (36.4 L) of water was employed. Secondary pressure grouting commenced 12 to 24 hours after primary grouting. Similar to the test anchors, a grout take of about 9 cu ft (0.25 m^3) per valve was strived for and generally achieved within several trials.

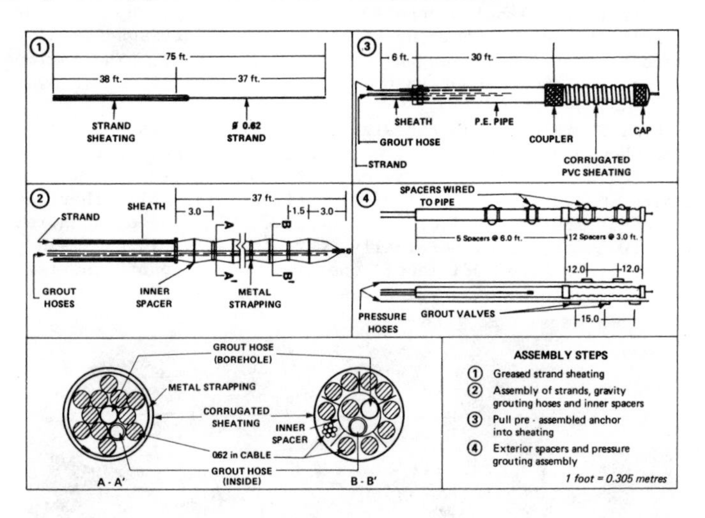

FIG. 12 - TYPICAL 320 K (1424 kN) MULTI-STRAND ANCHOR - ASSEMBLY DETAIL

ANCHOR ACCEPTANCE TESTING - Short-term and extended acceptance testing of all anchors was specified. The short-term tests consisted of loading to 0.1, 0.5, 1.0, and 1.4 of the working load, P_W. At a load of 1.4 P_W, displacements were measured at 1, 2, 5, 10, 15, 20, 25 and 30 minutes. On the displacement-log time graph, a creep rate criterion of less than 0.05 in (1.2 mm) per log cycle, had to be satisfied. Anchors were then unloaded to 0.1 P_W, to determine permanent displacement, reloaded to 1.4 P_W and then unloaded to the transfer load, which was usually 1.1 P_W to allow for some stress relaxation.

Extended acceptance tests, undertaken on 10% of the anchors, consisted of loading to 0.1, 0.4, 0.8, 1.0, and 1.4 P_W. Displacements were measured at 1, 2, 5, 10, 15 and 20 minutes for a load of 1.0 P_W, and additionally at 40, 50 and 60 minutes for a load of 1.4 P_W. Other test conditions and criteria were similar to the short-term test.

Very few anchors (4%) failed to meet the acceptance criteria; those that did were subjected to additional pressure grouting and retested. All passed the test requirements after this procedure. The initial anchor test program resulted in economies in final design and facilitated the formulation of site-specific acceptance criteria that, while rigorous, proved to be entirely workable.

MONITORING OF WALL PERFORMANCE

To monitor performance of the wall/anchor system, instruments were installed at critical locations, as shown on Fig. 13, prior to and during construction. These instruments consisted of survey plugs, tiltmeters, heave gauges, pneumatic piezometers and anchor load cells.

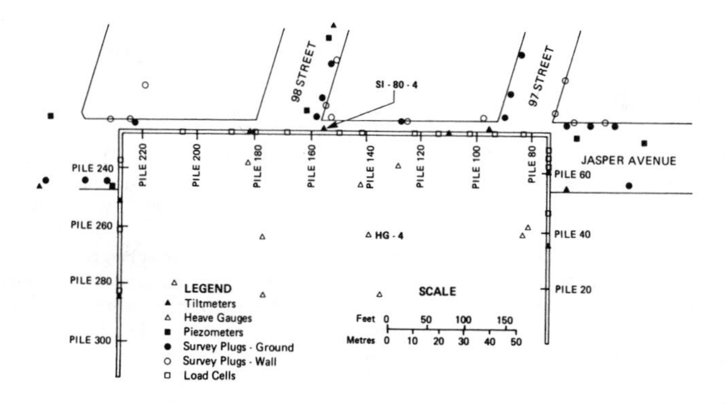

FIG. 13 - CONVENTION CENTRE INSTRUMENTATION

SURVEY MONUMENTS - Survey plugs adjacent to the northeast corner indicated total heave displacements of about 0.2 in (5 mm); diminishing to zero 80 ft (24 m) east of the corner. Back of the north wall, heave of about 0.8 to 1.2 in (20 to 30 mm) was measured. This heave diminished to about 0.4 in (10 mm) 80 ft (24 m) from the wall. Heave adjacent to the northwest corner was about 0.6 in (15 mm), and diminished to zero 50 ft (15 m) west of the wall. Although heave movements are attributed in part to elastic rebound, a substantial amount is due to pressure grouting of upper anchors.

Measurements on the north wall were also performed, as shown on Fig. 14. Allowing for variability in the readings, the maximum heave appears to have been about 0.5 in (13 mm) near the centre of the wall and near zero at the corners.

From Fig. 14, it is also apparent that the top of the north wall has deflected away from the excavation. This deflection is more pronounced near the east end where about 0.8 in (20 mm) of movement has been measured. The western half of the wall appears to be at or near its initial position.

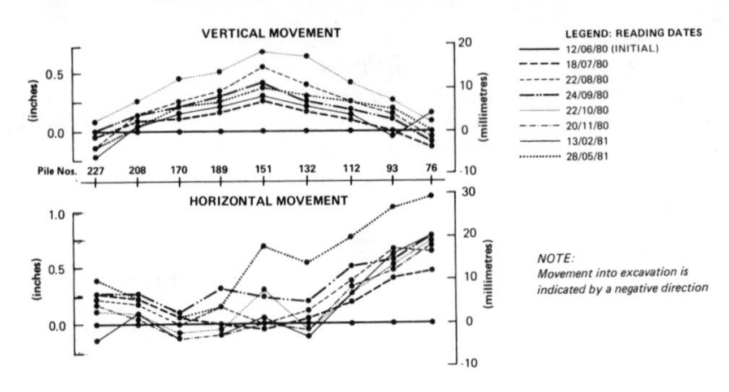

FIG. 14 - NORTH TANGENT PILE WALL DISPLACEMENTS

East and west walls have also experienced heave and lateral displacements. Observed heave has been about 0.2 in (5 mm) for both walls. Lateral displacement of the walls has been about 0.4 in (10 mm). Whereas the east wall has moved away from the excavation, the west wall appears to have moved toward it.

TILTMETERS - All tiltmeters, installed in and outside the tangent pile wall, have revealed some movements. The most significant movements have been recorded at substantial depth near the northeast and northwest corners, and near the centre of the north wall. The largest downslope movements were recorded, as shown in Fig. 15, at a depth interval of 150 to 154 ft (45.7 to 46.9 m). As evident from Fig. 15, this zone of movement, incorporates a layer of weak bentonite. It also corresponds approximately to the lower shear zone shown on Figs. 2 and 10. Three other tiltmeters revealed lesser movements at an approximately comparable elevation.

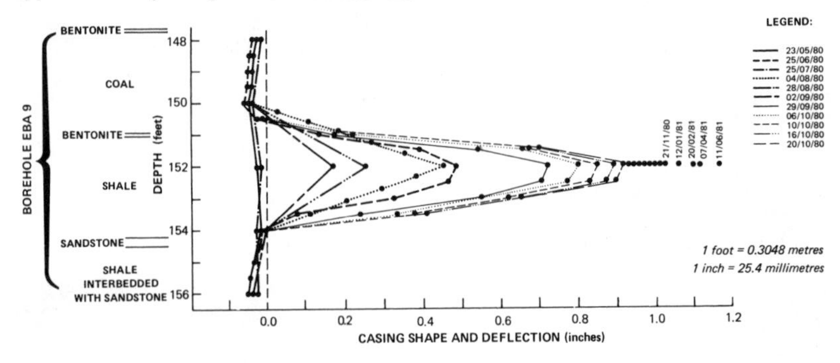

FIG. 15 - TILTMETER DISPLACEMENT BELOW CENTRE OF NORTH WALL

Because of concern over the possibility of a strain-induced massive landslide, movements were carefully monitored with time and the excavation process. Fig. 16 shows that a linear relationship existed between excavation volume and deflection at the critical depth. Hence, it was concluded that the movements were of an elastic nature and should slow down after excavation completion. Fig. 17 demonstrates that in October 1980, when the sixth excavation level was completed, movements slowed substantially. Although movements are still occurring, they are small and are anticipated to cease entirely in time.

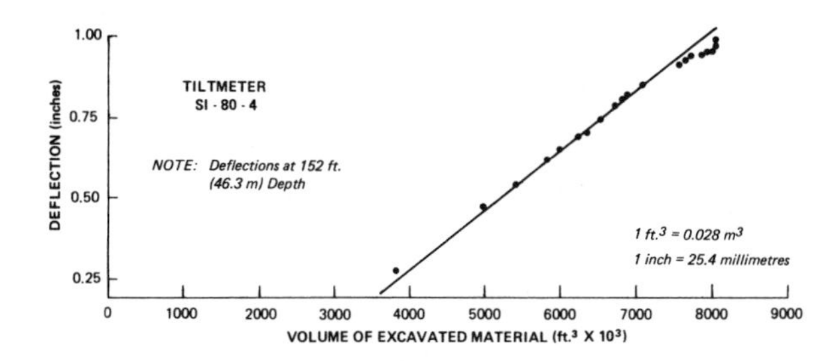

FIG. 16 - TILTMETER DEFLECTION VS. EXCAVATION

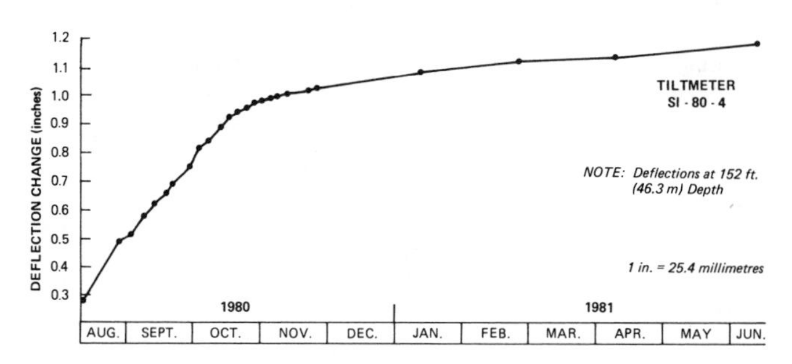

FIG. 17 - TILTMETER DEFLECTION WITH TIME

HEAVE GAUGES - Multipoint, ring magnet, heave gauges were installed in boreholes to depths of between 100 and 165 ft (30.5 and 50.3 m) below original grade. A sensor probe was used to measure magnet location relative to the excavation level, as shown on Fig. 18.

Unfortunately, construction activity resulted in a high rate of disturbance and instrument loss. The last set of readings obtained revealed a maximum base heave of approximately 4.3 in (110 mm) at the approximate centre of the excavation. At that time, the majority of the excavation had been completed. Greater base heave likely has occurred, but has not been measured. The response of heave to excavation was generally elastic. While consistent with observations of heave made elsewhere in Edmonton, it was greater than calculated.

PIEZOMETERS - Pneumatic piezometers were installed, prior to construction, to check the nature and level of groundwater response to construction activity. Periodic readings have revealed no change relative to pre-construction piezometric conditions.

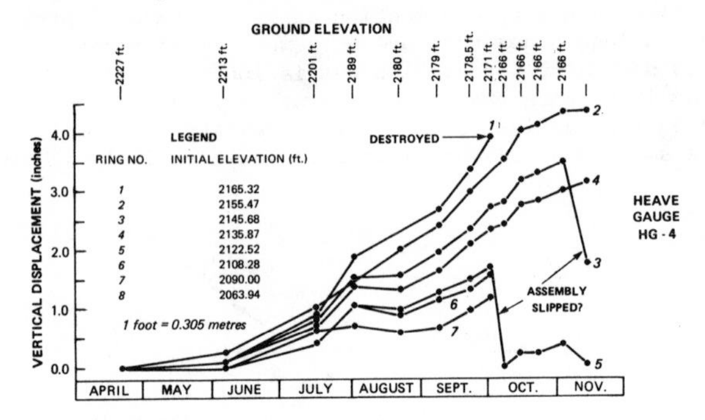

FIG. 18 - EXCAVATION BASE HEAVE - EDMONTON CONVENTION CENTRE

LOAD CELLS - Installation of load cells was specified on 5% of the anchors. The load cells, supplied by the contractor, were GLOTZL, hydraulic, direct reading units. The accuracy of this type of cell was checked prior to installation and approved.

Fig. 19 presents typical load cell readings with time. From this figure, it is evident that the Row 1 anchor loads fell off with time, and they had to be re-stressed several months after initial tensioning. This initial load loss was attributed primarily to stressing of anchors in lower rows. However, recent load loss is believed to be due to relaxation of the softer glacio-lacustrine soils behind the wall at the level of the upper two rows of anchors. No appreciable load loss has occurred in Row 4 anchors. This is typical of all anchors below the top two rows. As the anchors have remained exposed, the temperature sensitivity is again apparent in the plot.

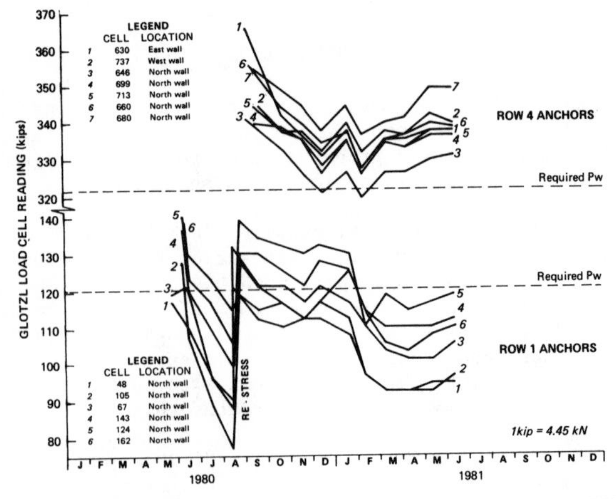

FIG. 19 - LOAD CELL READINGS - ROW 1 AND ROW 4 ANCHORS

CONCLUSIONS

Wall/anchor system performance has generally been as anticipated. Although instrumentation has indicated elastic movements greater than estimated from analyses, zones of movement were predicted satisfactorily. The deep shear zones, a major cause for concern and the reason for the comprehensive design, have proven to have been sensibly evaluated regarding their influence on site stability.

The anchor system performed well both during and after installation, with load levels remaining within reasonable limits of the design loads. The wall has correspondingly deflected to levels considered tolerable and consistent with the strain criteria believed essential for maintaining stability of the adjacent soil. As a consequence, the noted behaviour of the wall has provided the designers with a substantial level of confidence in utilization of permanent anchor systems in what are normally considered "weak ground" conditions.

ACKNOWLEDGEMENT

The cooperation of the Department of Real Estate and Housing and the Chief Commissioner's Office, of the City of Edmonton, in granting permission to publish this paper is gratefully acknowledged. The authors also extend thanks to their respective colleagues for contributions made to the analyses and designs utilized in the project.

APPENDIX - REFERENCES

1. Eigenbrod, K.D. and Morgenstern, N.R., "A Slide in Cretaceous Bedrock at Devon, Alberta", Geotechnical Practice for Stability in Open Pit Mining, ed. by C.O. Brawner and V. Milligan, 1976 pp. 223-238, AIME, New York
2. Hardy, R.M., and Associates Ltd., "Third Report re Greirson Hill, City of Edmonton, Alberta", a report submitted to the City of Edmonton in 1961 (unpublished)
3. Littlejohn, G.S. and Bruce, D.A., "Rock Anchors - State-of-the-Art", 1977, Foundation Publications Ltd., Brentwood, England
4. Locker, J.G., "The Petrographic and Engineering Properties of Fine-Grained Sedimentary Rocks of Central Alberta", Research Council of Alberta, Bulletin 30, 1969, 144 pages
5. Matheson, D.W. and Thomson, S., "Geological Implications of Valley Rebound", Can. Journal of Earth Sciences, Vol. 10, 1973, pp. 961-978
6. Ostermayer, H., "Construction Carrying Behaviour and Creep Characteristics of Ground Anchors", Preprint, 1974, Conference on Diaphragm Walls and Anchorages, London, England
7. Thomson, S., "Riverbank Stability Study at the University of Alberta, Edmonton, Alberta". Can. Geotech. Journal, Vol. 7, 1970, pp. 157-168
8. Thomson, S., "The Lesueur Land Slide, a Failure in Upper Cretaceous Clay Shale", Proc. 9th Conf. on Eng. Geol. and Soil Mech., Boise, Idaho, 1974, pp. 257-287
9. Thomson, S. and Yacyshyn, R., "Slope Stability in the City of Edmonton", Can. Geotech. Journal, Vol. 14, No. 1, 1977, pp. 1-16

SUBJECT INDEX

Page numbers refer to first page of paper

AUTHOR INDEX

Page numbers refer to first page of paper